Practical Astronomy

DATE DUE

			Printed in USA

Hong Kong
Milan
Paris
Santa Clara
Singapore
Tokyo

Neil Bone, BSc
The Harepath, Mile End Lane, Apuldram, Chichester
West Sussex PO20 7DZ, UK

ISBN 1-85233-017-1 Springer-Verlag London Berlin Heidelberg

British Library Cataloguing in Publication Data
Bone, Neil
  Observing meteors, comets, supernovae and other transient
phenomena. –
  (Practical astronomy)
  1. Meteors   2. Comets   3. Supernovae
  I. Title
  523
ISBN 1852330171

Library of Congress Cataloging-in-Publication Data
Bone, Neil, 1959–
  Observing meteors, comets, supernovae : and other transient
phenomena / Neil Bone.
      p.   cm.—(Practical astronomy)
  Includes bibliographical references.
  ISBN 1-85233-017-1 (pbk.: alk. paper)
  1. Astronomy—Amateurs' manuals.   2. Meteors—Observations.
3. Comets—Observations. 3. Supernovae—Observations.   I. Title.
II. Series.
QB63.B66   1998   98–18112
523—dc21   CIP

Typeset by EXPO Holdings, Malaysia
Printed at the University Press, Cambridge
58/3830-543210   Printed on acid-free paper

Preface

In many respects the night sky presents a remarkably constant aspect, changing only gradually with the seasons and with the movements of the Moon and planets. An orderly Universe, manifested by the fixed stars in the celestial vault, lay at the heart of many ancient philosophical and religious beliefs. From time to time, however, for as long as people have watched the skies, short-lived events – on timescales from a few seconds to a matter of weeks – have been witnessed which bear testimony to the sky's inconstancy. "Guest stars" – *novae* – or "hairy stars" – *comets* – have in their turn terrified and fascinated generations of sky-watchers. In ancient times, such interlopers were often regarded as portents of ill-fortune. Other celestial displays, such as brilliant auroral storms, fireballs or active meteor showers, also found their way into historical records from pre-telescopic times.

For the modern amateur astronomer, the sky naturally retains its fascination. Many observers are happy to tick off the various nebulae and clusters of the "deep sky", very much part of the unchanging firmament as perceived by the ancients. Many others, however, gain the most enjoyment from the pursuit of the unexpected: the transient phenomena of my title. The deep sky will always be there another night, but only the alert and aware will be able to make the most of a short-lived comet apparition – like the glorious week in the spring of 1996 when Hyakutake was at its best – or be ready for the perhaps once-in-a-lifetime occurrence of a 30-minute meteor storm, as might be delivered by the Leonids as the 20th century draws to its close.

Some transient phenomena are predictable, and can be prepared for, like occultations or eclipses; others come without warning. The thrill of the chase leads a hard core of dedicated observers to the pursuit of the latter. Around the world, there are those whose greatest ambition is to be first to the discovery of a new nova in our Galaxy, or a supernova in an external galaxy. Such events erupt rapidly, being visible for only a few weeks before fading from view, and early detection is essential if professional astronomers and others are to obtain scientifically useful observations.

Across more or less the whole spectrum of amateur observing – from naked-eye meteor watching, through telescopic study of the planets and on to deep-sky imaging with CCD cameras – there are transient phenomena to look out for. It is my aim in this volume to convey some of the flavour of just what is out there to be seen and, hopefully, to encourage the more casually interested to take a more careful look. Those of us who pursue astronomy as a hobby do so for enjoyment. That enjoyment can be enhanced by being aware of, and alert to, the possible occasional occurrence, during otherwise routine observing, of transient astronomical phenomena.

Neil Bone
Chichester, 1998 May 25

Acknowledgements

Many of my astronomical friends and colleagues have helped in the assembly of this book. Given the range of topics covered, it was only natural that I would have to call on several observers, experts in their own fields of work, to provide appropriate photographs and drawings with which to illustrate the text. I must put on record my heartfelt thanks to all those who so quickly came forward with suitable material.

I am most grateful to Melvyn Taylor for drawing up some of his excellent field charts for the chapter on variable stars. Gary Poyner, Director of the BAA Variable Star Section, rapidly agreed to the use of the light curves shown in that chapter.

Andrew Elliott provided useful advice on occultations, and a photograph of the system he has used to such good effect to record meteors on video. Among other electronic imagers whose help has been invaluable, Mark Armstrong generously allowed me to use his supernova discovery images from only a few weeks before the completion of my main text, while Nick Quinn provided excellent close-up views of Comet Hale–Bopp's coma. Maurice Gavin also came forward promptly with his unique image from the recent series of mutual events among Jupiter's Galilean satellites. Dave Briggs provided a superb electronic series of Saturn being occulted by the Moon.

The more traditional means of recording – drawing and photography – are also well represented here. Rob Bullen is an excellent astronomical artist, and kindly, and at very short notice, came up with drawings from his logbook of Comet Hyakutake and the Shoemaker–Levy 9 impact scars on Jupiter. David Graham provided a superb drawing of Saturn's most recent major great white spot, and Richard McKim trawled the BAA Mars Section archives for suitable drawings of a Martian dust storm in progress. Richard and David also helped greatly by reading through the first drafts of the respective sections of Chapter 6.

Bruce Hardie allowed me to use his solar prominence sketches and a photograph of one of the largest sunspot groups of recent times. Dave Gavine provided a limb profile for the Moon derived from one of the most memorable nights in Scottish amateur astronomy in the 1970s, the occultation of Aldebaran in 1978 August. Martin Mobberley dug deep into his photographic past (like many others, he is now a committed CCD user!) for the photograph of Aristarchus used in the lunar chapter. Together with Nick James and Glyn Marsh, Martin also takes credit for the wonderful Hyakutake tail-disconnection image in Chapter 8. Pam Spence kindly provided total solar eclipse images from the Caribbean event of 1998 February.

All the above are thanked for allowing me to use their material, in many cases only a few weeks old, ensuring that what follows is at least topical!

Thanks are also due to John Watson at Springer for persuading me to expand my initial ideas for the book beyond just the prospects for the Leonid meteor shower. Once again, I am pleased that John Woodruff has applied to a book of mine his editing skills and his knowledge of astronomy.

As ever, my wife Gina has put up with chaos, piles of untidy – but meticulously ordered in a random sort of way – paperwork, and the frustrations of a husband learning to use a new computer in mid-stream. To her, as always, my thanks. Thanks also to five-year-old Miranda for *not* hiding Dad's disks this time round.

Contents

Chapter 1

Introduction

Astronomy is one of the few scientific fields in which the amateur participant, with no professional qualifications, can still make a valid contribution. While it is no longer the case – as it was in the era of the wealthy gentleman scientist in the 19th century – that amateur equipment may actually be superior to that in professional observatories, developments in the manufacture of instruments, in particular the CCD camera, mean that, for a price, the modern enthusiast can achieve results which were first being obtained at major research institutions only a few decades ago.

Between the golden age of visual observation at the eyepiece in the early 1800s, when amateur and professional had more or less the same equipment at their disposal, and the modern era, there was a parting of the ways. In the 20th century the rise of advanced photographic and spectrographic techniques as applied to astrophysics initially put this field of study well beyond amateur reach. As a result, many amateur observing programmes became much more focused on lunar and planetary work, where visual detection at the eyepiece remained preferable to photographic views blurred by atmospheric turbulence.

Some have questioned the continued value of such routine telescopic work now that the Moon and planets are being investigated directly by spacecraft. However, monitoring by spacecraft is far from continuous – unlike amateur observing programmes, the progress of which is determined less by budgetary constraints than by the enthusiasm of their participants. The rapid eruption of spots on Saturn or the onset of Martian dust storms is more likely to be detected first by dedicated amateur watchers than by spacecraft. Indeed, at the time of the Voyager missions to Jupiter, simultaneous amateur observations were sought in order to coordinate what the spacecraft recorded with the ground-based appearance of the planet.

Routine monitoring of phenomena in the Solar System remains very much the domain of amateurs, who continue to make observations which are useful in many areas of research. In addition to planetary phenomena, transient events such as the apparitions of comets attract a lot of amateur interest. Year-by-year monitoring of meteor showers by amateurs has become an essential source of data for professional studies of how the dust streams that produce the showers evolve over time.

The professional field of astrophysics, a logical outgrowth in the late 19th century from the emerging technology of photography, larger telescopes, and the understanding of what spectra could reveal about stellar processes, certainly became a more specialised area, well beyond most amateur workers through the first three-quarters of the 20th century. The arrival in amateur hands of better photographic emulsions and processing techniques and, more recently, CCD cameras has seen the divide between amateur and professional in stellar research begin to close again. It is now possible for the suitably equipped amateur to at least try to duplicate astrophysical work which was once the preserve of the professional, such as obtaining and interpreting the spectra of stars. Extramural, distance-learning and other courses allow the dedicated amateur to understand more closely how such studies are, and can be, performed.

Amateur observers can (and do) make a significant contribution to research by discovering novae and supernovae, and eruptive variable stars in outburst. The effective professional study of these events often depends on rapid reporting of discoveries made by amateur astronomers. Given suitable equipment, clear skies and dedication, amateurs have almost as much chance as their professional counterparts – some would say *colleagues* – of being first to a major discovery in these fields. Professional-quality observations of these outbursts immediately following their discovery are also within amateur reach: amateur astronomers can now make photometric and astrometric observations of high standard.

It is perhaps in the days or weeks following a major discovery, provided that the object in question falls within the grasp of the equipment available to them, that amateur astronomers get to experience most closely the real excitement of scientific research. When magnitude estimates of a newly erupted nova need to be made hourly, or a bright comet requires nightly – and night-long – observation, there will not be the time to make profound analysis on the spot. Likewise, during intense meteor activity, as might occur from the Leonid meteors in 1998 and 1999, all that can be done is simply to try to get down the counts as accurately as possible. Only *after* such events, once the initial burst of excitement has passed, comes the chance to sit back and properly analyse the data to see what the observations meant. Such is the general experience of the working research scientist.

Another aspect of good scientific practice to which amateur astronomers are introduced by observing transient phenomena is its collaborative nature. While a great deal of work is done by observers operating alone in the field, it is usually when this work is pooled with results from others that the fullest picture emerges. Astronomy should certainly be seen more as a collaborative than a competitive activity. Amateur observers are encouraged to contribute raw data to national organisations such as the British Astronomical Association (BAA), the Royal Astronomical Society of Canada (RASC) or the Association of Lunar and Planetary Observers (ALPO, in the United States). Each of these bodies has sections dedicated to particular aspects of observing, and provides valuable guidance on how best to study the various targets. Local astronomy clubs and societies are also a valuable resource for the amateur, providing contact with other observers who will usually be prepared to share advice and experience.

Ultimately, most amateurs pursue astronomy as something they *enjoy*, which is as it should be! Astronomy is such an accessible science that it can be enjoyed at many levels, from the most casual stargazing – learning the constellations, say, or picking off the brighter deep-sky objects – to more serious work which may prove to be of scientific value. The chapters that follow are intended to provide guidance for those who have already progressed beyond the entry level and are seeking fields of activity in which to hone their observational skills and further their interest.

Most observers maintain a log-book of their astro-nomical work, which builds up over the years into a personal and permanent record – a diary of their observing career. The most memorable entries in anyone's log are usually of rare – perhaps once-in-a-lifetime – transient astronomical phenomena.

Resources

Association of Lunar and Planetary Observers, PO Box 16131, San Fransisco, CA 94116, USA.

British Astronomical Association, Burlington House, Piccadilly, London W1V 9AG, UK.

Royal Astronomical Society of Canada, McLaughlin Planetarium, 100 Queens Park, Toronto, Ontario, Canada M5S 2C6.

Chapter 2

Meteors

Meteor observing is one of the most popular activities among amateur astronomers, not least because it requires a minimum of equipment for it to be carried out effectively. Naked-eye watches demand no more of the observer than willing and patience. Provided that the watch is of reasonable duration (preferably an hour, or multiples of an hour), and the observer takes care to record the sky transparency (as limiting magnitude) throughout it, the data gathered can be used to learn much about the behaviour of meteor streams. During a watch the observer concentrates on an area of sky to one side of a radiant which is expected to be active on the night in question, and records details of each meteor seen: principally time of appearance, type (whether shower member or sporadic), magnitude relative to background stars or planets, approximate position (constellation), and the presence and duration of any persistent ionisation train along the meteor's path. It is also useful to note any pronounced colour, or flaring and fragmentation in flight. Table 2.1 gives an example from the author's observing log, recording activity during the Leonids of 1995 – a time when the shower was becoming more active.

Results from meteor watches, which may be carried out alone or with other observers independently recording their own details, are collected by national bodies such as the Meteor Section of the British Astronomical Association (BAA), the American Meteor Society, and the Meteors Section of the Association of Lunar and Planetary Observers (ALPO). Results from many independent observers are pooled and processed

Table 2.1. Sample meteor observations

Date: 1995 November 17/18
Observer: N. Bone
Location: Apuldram, West Sussex. Latitude 50° 49.8′ N Longitude 0° 48.3′ W
Conditions: Clear   **Limiting Magnitude**: +6.0
Watch Interval: Start 01:30 UT End 02:30 UT Duration 1^h 00^m

UT	Magnitude	Type	Constellation	Train (s)	Notes
1:43	1	Leonid	Gem–Tau	1	Yellow-blue, long
1:51	3	Sporadic	Gem		White
1:54	−2	Leonid	Hya	3	Yellow, short
2:03	−1	Leonid	UMa	3	Yellow
2:08	5	Sporadic	Gem–Ori		White, very fast
2:13	0	Leonid	Per–Ari	2	Yellow, fast
2:21	4	Leonid	Hya		White
2:21	4	Sporadic	Mon		White, fast
2:21	4	Leonid	Tau		White, very fast
2:29	−2?	Leonid	UMa		Yellow, in peripheral vision
2:30	4	Leonid	Hya		White, fast

to determine a number of parameters of meteor activity, including the zenithal hourly rate (ZHR), and magnitude distributions for shower members relative to the background sporadic activity present at the same time. The ZHR is a theoretical value for shower activity, based on what an experienced observer might be expected to see with the radiant overhead in a perfectly clear sky; in practice, account has to be taken of the radiant's actual and changing elevation above the horizon, and of the haziness of the sky during the observing session (Bone, 1993; Bone, 1995).

Many basic astronomical texts give the impression that meteor showers behave in very much the same way from one year to the next. This is far from the case, however, and part of the fascination for many regular observers is the inherent unpredictability of showers, even the Perseids (Section 2.2) and the Geminids. During the 1970s the Geminids were described as having a sharp activity peak, with ZHR at best reaching about 60–70 (corresponding to typical observed rates from a reasonably dark location of about 40 meteors per hour). By the late 1990s this peak had become broad, with a period of over 24 hours at or above ZHR 70, and perhaps 6–8 hours when the ZHR was closer to 120. This change in behaviour is consistent with theoretical models of the parent meteoroid stream developed in the 1980s, and it will be interesting to see how closely the activity continues to follow predictions as the 21st century unfolds. The expectation is that the shower will start to subside as the stream's orbit is dragged away from the Earth, and by 2100 the Geminids may no longer brighten northern midwinter nights.

Meteor streams are by no means fixed in space, being subject to gravitational perturbations by the planets. These perturbations lead in turn to variable activity over the course of many years and, as with the Geminids and historical examples such as the November Bielids, may lead to a shower becoming inactive. Similarly, previously unidentified streams may be brought "on line" over time; indeed, the Geminids seem to have been unknown before the 19th century.

Many meteor streams have an inherent variability resulting from a continual input of new material from their parent comet. The Perseids are a prime example. Variations in the October Orionid shower from year to year are presumed to have their origin in the filamentary nature of the stream, different filaments having been laid down at separate returns to perihelion of the parent comet, 1P/Halley. Other, younger showers such as the Giacobinids and Leonids have their complement of meteoroidal material largely concentrated in one part of the stream's orbit, relatively close to the parent body, and substantial activity is thus found only in the years close to the comet's perihelion.

2.1 The Leonids: A Storm Coming?

Towards the end of the 20th century, a shift in the demography of meteor observing is likely! Many casual meteor observers restrict their efforts to the milder nights of August, close to the maximum of the Perseid shower, when good rates are more or less guaranteed and the weather is usually favourable. Those looking for a feast of high activity in the years around the turn of the century will doubtless put more effort into the nights around November 17 and 18, as the Leonids come to peak.

The Leonids are renowned for their periodic storms – outbursts of very high activity lasting less than an hour but during which the observed meteor rate climbs to thousands per minute, sometimes well beyond any level which can be reliably counted by a visual observer. Storms are possible in the years close to the perihelion of the parent comet, 55P/Tempel–Tuttle, at intervals of approximately 33 years.

Searches through historical records, including Korean and Chinese annals, reveal that the Leonids were active as early as AD 902, and possibly even as far back as AD 585. The first really well-documented Leonid storm was that witnessed from South America on 1799 November 11 by the explorers Alexander von Humboldt and Aimé Bonpland. The display was also visible, just before dawn, from the British Isles. Humboldt's detailed investigation of the phenomenon led to the discovery that such displays were periodic (an earlier storm had been seen from South America in 1766), and the forecast that a further storm would occur in the 1830s. The forecast was borne out by an immense Leonid storm on the night of 1833 November 12, visible from the West Indies and Canada, during which peak Leonid rates may have reached 200 000/h. A further repetition over western Europe in 1866 was intensely observed by scientists seeking to clarify the nature of meteors and their association with comets.

With the 33-year periodicity in Leonid storm activity established, many anticipated a further exceptional display in 1899. This, however, failed to materialise. The Leonids were reasonably active in the years around this perihelion return of 55P/Tempel–Tuttle, but the rates were a disappointment to those who expected to witness a spectacle to rival the displays of 1833 and 1866. A further "failure" in 1933 led to speculation that the shower had more or less petered out. When better methods for calculating orbital perturbations became available in the 1960s, it became apparent that the absence of storm activity in 1899 and 1933 actually stemmed from Jupiter and Saturn having gravitationally perturbed the core parts of the meteor stream away from the Earth's orbit so that we no longer encountered the main concentration of meteoroids, called the ortho-Leonid cloud, close to the comet itself. Further perturbations, however, brought this part of the stream close to us again for the mid-1960s, and Harold Ridley, then Director of the BAA Meteor Section, made the confident prediction that a storm would again be likely in 1966.

Ridley's prediction came true on the night of 1966 November 16/17. Observers in western Europe saw only a modest return (though still better than an "average" Perseid display), but for a period of about

40 minutes those in the western United States were treated to the meteoric show of a lifetime. Like previous Leonid storms, that of 1966 was short-lived, and so visible only in a restricted longitude zone of the Earth. Leonid rates of 160 000/h were estimated for the 40 minutes of highest activity.

It would be wrong to give the impression that activity was at storm levels throughout the night on the occasions when the Leonids produced their most spectacular displays. The storms are short-lived spells of very intense activity superimposed on the "background" Leonid level, which is substantial (as already mentioned, better than the Perseids' regular activity) in the years around 55P/Tempel–Tuttle's perihelion. The short duration of storm activity restricts an individual observer's chances of witnessing it: much effort has been put into forecasting just where any storm peak might lie, in longitude, around 1998/99.

2.1.1 The Quiet-time Leonids, 1970–1993

Leonid activity is far from absent in the years away from the perihelion of 55P/Tempel–Tuttle. Dedicated meteor observers have continued to monitor the shower over the years, finding activity at least as good as, if not better than, the other major shower of November, the Taurids. The Leonids are presently active between about November 15 and 20. The peak occurs around November 17 or 18, the precise date depending on whether the year in question is, or is close to, a leap year. A useful convention for the study of meteor showers, which avoids any complications introduced by leap years, is the system of describing the Earth's orbital position in terms of solar longitude, adjusted to the standard epoch of 2000.0. Since one orbit around the Sun takes 365.25 days, we arrive back at the same solar longitude 6 hours later, in Universal Time, each year. Thus, in identifying the likely position of any great Leonid return at the end of the 20th century, it is useful to consider the position of the maximum in solar longitude, and back-calculate from this to UT in order to find the geographical locations likely to be in darkness at the time of the peak and with the radiant above the horizon.

The Leonid radiant lies in the Sickle of Leo, which in mid-northern latitudes rises out of the eastern haze of a late autumn evening around 23h local time. The radiant reaches its highest elevation above the horizon in the pre-dawn hours, and the best observed rates will generally be found after about 03h local time. There is no point in looking for activity before the Leonid radiant has risen (as many in the UK seem to have done, to their disappointment, on the night of the storm in 1966).

Following the 1965 perihelion of 55P/Tempel–Tuttle and the 1966 storm, substantial Leonid activity was seen up to 1969, when good numbers were reported by observers in North America. The returns of 1972 and 1974 also produced respectable if more modest activity. Thereafter, rates more or less returned to their "quiet-time" levels. Figure 2.1 shows the activity profile, as a function of solar longitude, derived from Leonid observations submitted to the BAA Meteor Section between 1981 and 1993. Activity in each of these years was fairly similar, with generally low observed rates of 3–5/h in the first few days of the shower, corresponding to a corrected ZHR of around 5–10. A pronounced peak is evident around solar longitude 234.5°, with ZHR in excess of 15, after which activity subsides more slowly until about November 20.

Figure 2.1. ZHR profile for the quiet-time Leonids, derived from BAA Meteor Section results obtained between 1981 and 1993.

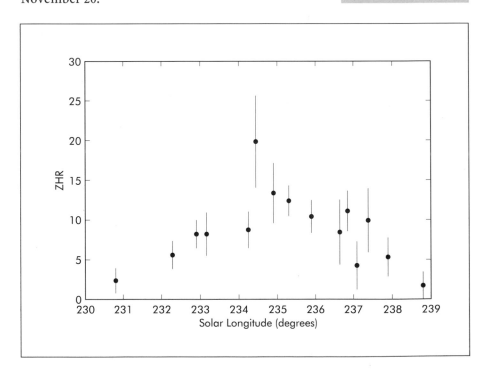

Leonid meteors are notably fast, with a geocentric velocity of typically 70 km s^{-1}. As a consequence the shower has a high proportion of meteors showing a persistent train, produced by ionisation along their path, which remains luminous for several seconds after the meteor itself has extinguished. This remained a prominent feature of the Leonids through their quiet-time activity. Also of note was the continued abundance, relative to the sporadic background, of bright events.

2.1.2 On the Rise: 1994–1996

Seasoned meteor observers kept a close eye on the Leonids through the early 1990s, finding nothing unusual up to 1993. Bright moonlight was a problem in 1994, swamping the fainter meteors. Despite this, several observers in Spain and the United States were able to detect Leonid activity on the rise, with observed rates – even in the lunar glare – of about 20–30/h, well above the quiet-time levels. This rising activity, some four or five years ahead of the anticipated storm peak(s) in 1998–99, was very much in line with a similar rise witnessed before the previous perihelion return of 55P/Tempel–Tuttle, in 1961.

In the light of these observations, the relatively moonless return of 1995 was awaited with considerable excitement. For once, observers in western Europe had excellent weather conditions. A tongue of cold, clear Arctic high-pressure air, extending all the way down from Greenland, covered the British Isles, giving transparent skies which were well exploited by meteor observers. The BAA Meteor Section received its largest collection of Leonid data for a great many years from the night of maximum, November 17/18. The derived ZHR plot (Figure 2.2, *overleaf*) and magnitude distribution (Figure 2.3, *overleaf*) are shown here. Early in the night, while the radiant was low, observed rates were comparatively modest. After about 02h local time, however, activity became more substantial. For a period of about 90 minutes around 04:00 UT (solar longitude 235.4°) rates were particularly high, suggesting a peak ZHR of about 40 – more than twice the "quiet-time" level. Similar activity, with observed rates of up to 30/h, was reported by other European groups at the same time.

Not only was Leonid activity higher than most observers had previously enjoyed in many hours of

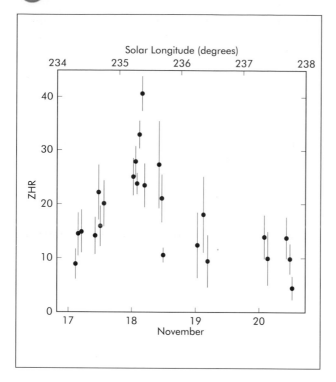

Figure 2.2. Leonid ZHR profile for 1995, based on BAA Meteor Section results. A marked enhancement in activity is evident in the early morning of November 18, around solar longitude 235.4°.

Figure 2.3. During their 1995 return the Leonids showed a marked abundance of bright meteors relative to the sporadic background.

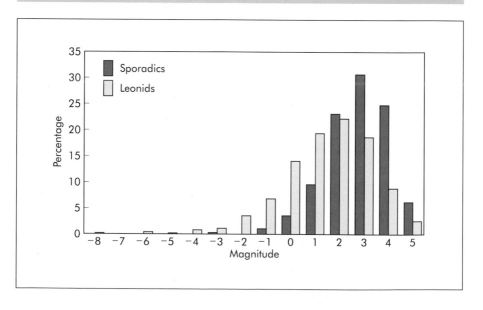

November meteor watching, but also the numbers of bright events, and events with persistent trains, were high. Several negative-magnitude Leonids were photographed, and one magnitude –5 event, seen well from locations in the southern and western British Isles at 04:36 UT, left a persistent train lasting for up to 5 minutes.

Weather conditions were less favourable for British observers in 1996, but elsewhere in Europe skies were rather better. Limited observations in the pre-dawn from the UK on 1996 November 16 showed minimal Leonid activity, but within a few hours more substantial Leonid rates were being recorded from Canada, with ZHR in the mid-20s.

Only a handful of UK-based observers, in Northern Ireland and Ayrshire, had clear skies on November 16/17. Good activity was recorded behind a clearing cold front, with typically about 30 Leonids per hour, translating to a ZHR of 40–50. As in the previous year, several spectacular bright Leonids were noted, many leaving long-duration persistent trains. Notable among these, a mag. –5 event at 05:32 UT, seen by Terry Moseley and Nick Martin on either side of the Irish Sea, left a train lasting for up to 16 minutes!

High activity had been evident as early as 00:00 UT, as reported by Dutch observers fortunate enough to find a gap in the clouds over France. The elevated rates continued after dawn had broken over Europe, as shown by results obtained by James N. Smith in Canada. Up to about 09:00 UT (solar longitude 235.3°), Leonid ZHRs were around 60–70. Results communicated by Robert Lunsford in California indicate that the ZHR remained as high as 50 until 13:00 UT – the peak of Leonid activity substantially above quiet-time levels had, as in the early 1960s, broadened into a "plateau" of some 12 hours (Figure 2.4, *overleaf*). Unlike the 1995 return, there was no obvious single sharp peak suggestive of a particularly dense filament in the Leonid stream; the 1996 Leonids offered no real clues as to the likely position of any storm peak later in the 1990s.

Bright Leonids were abundant relative to the sporadic background in 1996, as reflected by respective mean magnitudes of +1.29 and +3.07. Particularly for events of mag. +1 and brighter, the Leonids showed a marked excess over the sporadic background. Persistent trains were reported for 50.0% of Leonids, compared with 14.7% for sporadics.

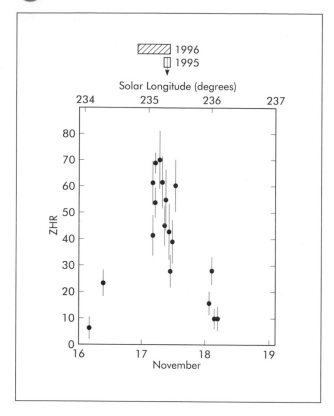

Figure 2.4. Enhanced Leonid activity continued in 1996, with a broader time-span of rates above quiet-time levels. As indicated by the bars at the top of the ZHR profile, activity was above quiet-time levels for 12 hours in 1996, compared with 90 minutes in 1995.

2.1.3 The Possible Storm Years: 1997–2000

Comet 55P/Tempel–Tuttle was recovered as a very faint object in early 1997 March, on much the expected orbit. More precise measurements of its position and motion are expected to refine the likely solar longitude of the Leonid peak in 1998–99. Perihelion was reached on 1998 February 28, close to the point at which the comet crosses the descending node of its orbit with respect to the Earth's orbital plane.

Unfortunately, a waning gibbous Moon rising a couple of hours ahead of the Leonid radiant seriously interfered with visual observations in 1997, when the Earth passed the node 107 days ahead of the comet. Extensive studies by John Mason and Joe Rao have shown that storm activity can happen this far ahead of the comet, depending on its proximity to the stream's orbit. For example, the 1766 and 1799 storms were seen when the Earth encountered material similarly far ahead of the comet at the node; at the 1799 return, the Earth was much closer to the stream's orbit than in 1997.

Analysis of the historical records suggests that most Leonid storms occur when we pass through the node *behind* the comet, at times when the comet's orbit lies slightly inside that of the Earth; 1998 and 1999 are such occasions.

Much as expected, the roughly 12-hour period of highest activity was missed from western European longitudes, where only low Leonid rates were observed. Since we arrive at the same solar longitude 6 hours later each year, the region of enhanced Leonid activity is encountered from locations correspondingly farther west on the globe, occurring in 1997 from about 06:00 to 18:00 UT on November 17. Observers in the western United States, and on Hawaii, were particularly favoured. Again, despite the moonlight excellent Leonid activity (but no storm) was reported, with numerous bright meteors and persistent trains. Observed rates of 55–60/h were reported from California around 12:00 UT (solar longitude 235.22°); moonlight makes a reliable estimate of corrected ZHR difficult, but a value of around, or slightly in excess of, 100 seems reasonable. This level of activity was still in evidence a couple of hours later for observers on Hawaii.

The prospects for Leonid meteor storms in 1998 and 1999 are excellent, with the Earth crossing the node of 55P/Tempel–Tuttle's orbit 258 and 623 days, respectively, after the comet. Intuitively, we might assume that the 1998 passage, being closer behind the comet, is the likelier to produce high Leonid rates. By analogy with the comparable return in the 1960s, however, the later, 1999 passage may prove the more productive; the 1966 storm occurred when the Earth passed the node some 561 days after the comet. Meteoroids appear to be distributed most densely in the ortho-Leonid cloud outside the comet's orbit at around this distance. A substantial display, though still perhaps less intense than that of 1966, was recorded by radar observations in 1965 some 196 days behind the comet. Visual observations in 1965 were restricted, however, by a strong, broad waning crescent Moon, 24 days old and rising soon after the radiant at the time of maximum.

Forecasting the likely time of a storm peak in 1998 and 1999 remains difficult, especially since there was no indication of a strong isolated peak superimposed on the 12-hour span of enhanced rates seen in 1996 and 1997. Moonlight may have hidden any short-lived visual peak of faint meteors (of which the storm activity is largely comprised) in 1997, but there is nothing in the available radio data, either, to indicate such a peak.

Taking the 1995 peak around solar longitude 235.3–235.4° as a reasonable possibility, we might expect to encounter the densest parts of the ortho-Leonid cloud around 21:00–22:00 UT on 1998 November 17/18, and around 03:00–04:00 UT on 1999 November 17/18, consistent with the time of closest passage to the current longitude (235.26°) of the node of 55P/Tempel–Tuttle's orbit. Observers in the Far East and, possibly, eastern Europe are most likely to be favoured in 1998, though those in western Europe may catch a glimpse of some of the peak activity as the radiant rises. In 1999, western Europe will be under the best of the Leonid display, with the radiant high in the south-eastern sky. These times are, however, subject to considerable uncertainty – the peak may be several hours earlier or later.

To either side of any storm peak in those years, observers can expect to see several hours of Leonid activity comparable to that recorded in 1997: observed rates well in excess of 50/h, including bright meteors with persistent trains. In 1998 observers in Europe and possibly the eastern seaboard of North America will be best placed for the 12-hour plateau on November 17/18. In 1999, western Europe and the whole of North America will be favoured.

The Leonid peaks in both 1998 and 1999 are blessed with an absence of moonlight. In 1998 the Moon is new, and will not interfere with observations, while the first quarter Moon will be setting as the radiant rises in 1999.

A significant Leonid display is also quite likely in 2000, though perhaps not to storm levels. Observers, in North America particularly, should watch the Leonids closely in this year, despite the presence of a waning last quarter Moon in the post-midnight sky.

2.1.4 Observing the Leonids

Whatever happens in the years surrounding the return of 55P/Tempel–Tuttle, it is critically important that amateur observers, as well as others interested in meteors, make as much use as possible of any observing opportunities. November nights at mid-northern latitudes are often cloudy, unfortunately, and many meteor enthusiasts are seriously considering travelling to more favourable locations for the Leonid returns of 1998 and 1999. Such "astro-tourism" – normally associated with eclipse-chasing – has been a major impetus behind attempts to forecast the timing of the Leonid peaks in these years. In the coming years observers should make use of whatever clear sky they can find during the entire span of Leonid activity, from November 15 to 20, in order to obtain as full a record as possible.

Clear November nights are cold, and would-be observers are advised to wrap up warmly against the elements. Undoubtedly, most observers will wish to carry out naked-eye watches. The important thing to remember is that such watches are of scientific use only if the details are carefully and accurately recorded. Pay particular attention to the state of the sky during watches, and make a note of the limiting magnitude; hazy conditions will adversely affect the numbers of meteors recorded, particularly faint ones. Accurate timing is also important, so ensure, before starting to observe, that your timepiece is synchronised to a standard time signal.

For most of the time, when rates are comparatively steady, it should be possible to record meteors in the standard way outlined earlier – time, type and magnitude being the principal details to note. A record of the sporadic activity at the time of the watch is important: record *all* meteors seen, not just Leonids.

Should activity start to climb markedly, it will be prudent to abandon the recording of each meteor in detail and change to a system of counting in blocks, of perhaps 5 minutes at a time, or shorter intervals if activity starts to get really hectic. During a full-blown meteor storm, with meteors appearing all over the sky at rates of several per second, the best approach may be to count those appearing in a limited area of sky such as the bowl of the Plough or the Square of Pegasus. The aesthete might, at this point, advocate simply sitting back and enjoying the spectacle, since the accuracy of any attempted count may prove low!

Many experienced observers prefer to record their observations using the traditional media of pen and paper, often adopting a shorthand code to get the details down quickly and minimise the dead time spent not looking at the sky. Some use pocket tape recorders, which are all well and good, provided they can be protected from the cold and damp that can lead to battery or equipment failure and consequent loss of valuable data. For rapidly changing high activity, some sort of tape record, encoded with time-marking, may be advantageous.

During very high activity, other forms of recording come into their own. Photographers should be able to record the brighter meteors, using driven or undriven time exposures with a reasonably fast lens ($f/2.8$ or lower) having a fairly wide field of view. Exposures of 10–20 minutes' duration on a suitable emulsion (the BAA Meteor Section recommends Ilford HP5plus) may succeed in recording meteors brighter than mag. 0. For periods of intense activity, photography offers the chance of getting some sort of count for the brighter meteors if exposures are kept to 1 minute or less, and the start and finish times for each frame are accurately recorded.

Perhaps the most promising instrumental advance for meteor work to have come into amateur hands since the 1966 Leonid storm is the low-light video camera, with which it is possible to record activity in real time for later more precise counting. A number of amateurs routinely use video to record major shower activity, notably the German observer Sirko Molau, and Andrew Elliott and John Mason in England. Andrew Elliott's system, operated at Reading, employs a 28-mm wide-angle lens giving a 50° field of view to a low-light camera, images from which are recorded on videotape (Figure 2.5). A time-stamp

Figure 2.5. The low-light video system used by Reading amateur Andrew Elliott to record meteors.

is simultaneously recorded on the tape. The image is updated 25 times per second, offering a timing accuracy of 0.04 seconds. On playback it is possible to review individual frames, so with such a system the intensity of any Leonid storm activity can be accurately estimated after the event.

The transient and unpredictable nature of putative Leonid storm activity in the years around the end of the 20th century is likely to attract an enormous amount of interest from observers. Calculations of the future evolution of the stream's orbit suggest that this may be our last chance to see a Leonid storm for many centuries; in 2032, the miss-distance between Earth and the core of the ortho-Leonid cloud will be too great for very high activity, though a respectable shower should still be seen.

2.2 The Perseids: The Unexpected Rise of "Old Faithful"

A regular highlight of the year, even for observers who would otherwise profess no interest in meteors, is the Perseid shower, active from July 25 until about August 20. Peak is reached around August 12, and the dependably high rates – typically one meteor per minute close to maximum – and mild observing conditions make the shower a great favourite. The shower's dependability has led some to dub it "Old Faithful". Events during the late 1980s, and into the 1990s, have shown that even the apparently steady Perseids can spring the occasional surprise.

Figure 2.6 (*overleaf*) shows the behaviour of the Perseids in a fairly typical year, based on BAA observations from 1983. Rates are low into the first few days of August, showing a slow, gradual rise until about August 10, after which activity becomes markedly more substantial. The peak itself, around solar longitude 140.0°, is fairly short-lived, with the period of best rates lasting 6–8 hours. In an average year the peak ZHR is usually about 80, though some variations have been seen from year to year. There are suggestions that the shower became more intense at peak through the 1970s; a very high return, with a peak ZHR of 120, occurred in 1980. The 1980 return was particularly well observed from the British Isles.

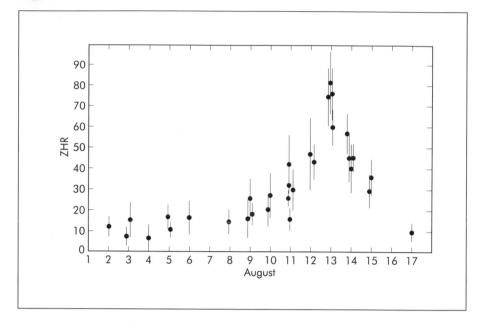

At this time the search was on for the expected return of the Perseid meteor stream's parent comet, 109P/Swift–Tuttle, which had previously been seen in 1862, and was thought to have an orbital period of about 120 years. It seemed quite reasonable to assume that the high Perseid peak of 1980 was a result of an enriched cloud of debris in the meteor stream close to the as yet unseen comet. A recurrence of this high activity in 1981 over the United States led to the suggestion that 109P/Swift–Tuttle had, indeed, returned to perihelion, but had somehow managed to slip past undetected, perhaps in the Sun's glare. Further weight was given to this idea by the shower's apparent reversion in 1983 to more normal peak levels.

The plot thickened, however, in the late 1980s, when analysis of observations by the fledgling International Meteor Organization led to the realisation that in 1988 the Perseids had shown two more or less equally strong maxima, separated by about 12 hours. The regular maximum at solar longitude 140.0° had been preceded by a new, previously unseen feature in the activity profile.

Unfavourable moonlight conditions limited Perseid observations in 1990, but at the next favourable return in 1991 the new peak was again evident, now some 11 hours ahead of the regular maximum and showing a much greater intensity. Observations from the Far East

Figure 2.6. Perseid ZHR profile from BAA Meteor Section results in 1983. This is a relatively "normal" return, with maximum on August 12/13.

indicated high activity for an interval of about an hour around solar longitude 139.58°, during which time the ZHR reached about 350 for short intervals. The ZHR estimates from this time were based on short-interval counts, extrapolating to an *equivalent* hourly count, and might therefore be referred to more correctly as "EZHR" estimates. Western European observers saw the regular maximum at solar longitude 140.0° with a ZHR of 80, much as normal.

The 1992 return was also badly affected by moonlight. Once again, however, visual observers in eastern Europe reported a period of high activity around solar longitude 139.5°, with postulated EZHR of about 240.

A very useful adjunct to visual and photographic meteor work is provided by forward-scatter radio observations. A short-wave receiver is tuned to a transmitter over the horizon whose signals therefore cannot normally reach the observing station. Reflection of radio waves from the transient column of ionisation left by a meteor high in the atmosphere above the area between transmitter and receiver leads to the reception of a short burst of signal. These "pings" can be counted in order to assess meteor activity during daylight or under cloudy conditions, provided there is no other atmospheric interference. Forward-scatter radio results of this kind obtained from England by John Mason clearly showed a sudden onset of enhanced Perseid activity on the evening of 1992 August 11/12, commencing around 18:45 UT at solar longitude 139.42°. The outburst, as detected by radio, consisted of short bursts of high activity separated by brief quieter interludes, indicative of separate filaments of material in the stream being encountered by the Earth at the time.

Clearly, something unusual was going on in the Perseid meteor stream, and it was not long before an explanation was forthcoming. Comet 109P/Swift–Tuttle was eventually recovered, ten years later than anticipated, in the autumn of 1992. The comet nucleus was found at this perihelion return to be particularly active, with many jets of material emerging into the coma; it seems likely, therefore, that the comet is subject to marked non-gravitational perturbations in its orbit, causing it to return on this occasion much later than expected. The orbital period has been refined from the 1860s value of 120 years to 130 years, and the next return, in 2126, is expected to be spectacular, with the comet passing close to the Earth.

The parent comet's recovery, together with parallels with the Leonids (the Earth would, in 1993 August, pass through the descending node of the comet's orbit some 224 days behind 109P/Swift–Tuttle itself) fuelled predictions of a possible Perseid meteor storm around the solar longitude of the newly emergent early peak. Any such storm might have been expected to occur around dawn in western Europe on the morning of August 12. A great deal of speculation found its way into the popular media, much of it ill-informed and leading to exaggerated public expectations. In the event, weather conditions over much of western Europe proved very poor, and very few observers managed a glimpse of the shower around the critical time.

Where observations were possible, in France, parts of Scotland, and the north and east of England, observers found higher than expected Perseid rates throughout the night, with a Leonid-type plateau of some 3 hours during which rates particularly high, but not at storm levels (Bone and Evans, 1996). The ZHR was found to be about 70–80; in a normal Perseid return the ZHR around this time would be closer to 30–40. Superimposed on the plateau at around 02:00 UT was an interval of some 30 minutes with much higher activity, around solar longitude 139.47°, with EZHR around 180 (Figure 2.7). As dawn was beginning to break over the British Isles, rates declined to the earlier plateau level. The regular maximum, occurring over the Far East, again reached a ZHR of 80, and was clearly unaffected by the recent perihelion passage of the parent comet.

The 1994 return brought the possibility of a recurrence of the high, early peak over North American longitudes, and observers in the United States and Canada were not disappointed. The early peak again showed itself around solar longitude 139.58°, superimposed on a 3–4 hour plateau of enhanced rates. Over Europe, the regular maximum was much as in the previous year (Bone and Evans, 1997), reaching a ZHR of about 80.

Further demonstrating that meteor activity can be unpredictable, the Perseids continued to show unusual activity well "downstream" of the parent comet. While the 1995 return was afflicted by a full Moon, there was no such interference in 1996, and observers in western Europe again recorded unusually high rates ahead of the regular maximum. The additional peak, now around solar longitude 139.66°, gave a ZHR of around 90, much diminished from its 1993–94 levels but still

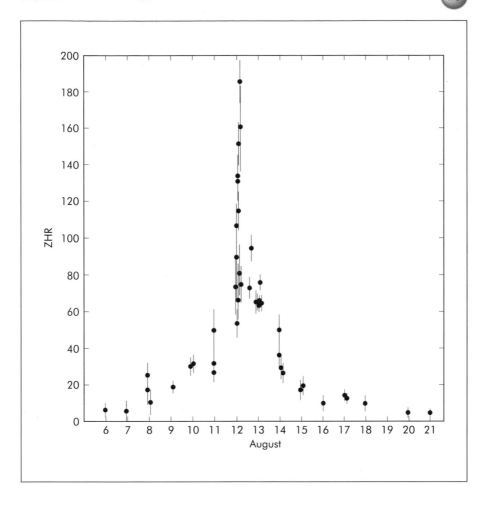

Figure 2.7.
Following the perihelion of the stream's parent comet 109P/Swift–Tuttle in 1992, the Perseids of 1993 showed an unusual extra, high, sharp activity peak 12 hours ahead of the regular August 12/13 maximum.

markedly higher than Perseid activity at the equivalent solar longitude in a normal year (Figure 2.8, *overleaf*). A further repetition in 1997, seen from North America, brought activity at a similar level around solar longitude 139.72°. With the passage of time since the major returns of 1993–94, the additional peak appears to be moving systematically closer to the established peak at solar longitude 140.0°, and the two will probably merge, or the early peak simply disappear altogether, early in the 21st century. That enhanced activity associated with the return of 109P/Swift–Tuttle should have continued for as much as 5 or 6 years after perihelion has been a considerable surprise to many, and emphasises the value of keeping showers under continual observation, even in the years away from expected outbursts. We may find similar surprises in the Leonids'

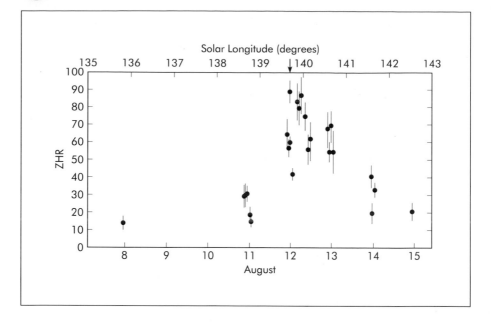

behaviour in the years after 2000; only observations will tell.

2.3 Other Showers in Outburst

The Leonids are not the only shower observed to have produced storm rates. Perhaps the best known are the Bielids (Andromedids), associated with the defunct comet 3D/Biela, which produced spectacular displays on November 27 in 1872 and 1885, and lesser outbursts in 1899 and 1904. The 1885 return is estimated to have produced Bielid rates of up to 75 000/h. Sadly, the stream's orbit currently passes far outside that of the Earth, having been perturbed outwards by the gravitational influence of Jupiter. While no activity is seen from the Bielids at present, computer modelling of the stream's orbit suggests that the shower may return, following further perturbations, around 2120.

Among other showers noted for their periodic outbursts (Table 2.2) are the Giacobinids, associated with Comet 21P/Giacobini–Zinner. The meteoroid cloud surrounding this comet is still very compact; unlike the situation with the Leonids, there appears to be no significant Giacobinid activity when the stream's parent

Figure 2.8.
Remarkably, the extra early peak in the Perseid ZHR profile has remained in evidence for several years following the perihelion of 109P/Swift–Tuttle. As indicated by the arrow in this 1996 ZHR profile from BAA Meteor Section results, the additional peak was diminished in intensity but nevertheless still in evidence. Similar activity was seen in 1997.

Table 2.2. Regular annual meteor showers and those subject to outbursts

Shower	Activity Dates	Maximum	Normal Peak ZHR	Radiant RA	Radiant DEC	Comments
Quadrantids	Jan. 1–6	Jan. 3–4	120	$15^h 28^m$	$50°$	Short, sharp maximum.
Lyrids	Apr. 19–25	Apr. 21	15	$18^h 00^m$	$32°$	Outbursts in 1803, 1922, 1982.
Pi Puppids	Apr. 15–28	Apr. 23	—	$07^h 20^m$	$-45°$	Best from S hemisphere. Outbursts 1977, 1982.
Eta Aquarids	Apr. 24–May 20	May 5	50	$22^h 20^m$	$-01°$	Best from S hemisphere.
Delta Aquarids (S)	July 15–Aug. 20	July 29	25	$22h 36m$	$-17°$	Best from southerly latitudes.
Perseids	July 25–Aug. 20	Aug. 12–13	80	$03^h 04^m$	$58°$	Outbursts between 1988–1997, strongest 1992–94.
Giacobinids	Oct. 7–10	Oct. 8	—	$17^h 23^m$	$57°$	Usually inactive. Outbursts 1933, 1946, 1985.
Orionids	Oct. 15–Nov 2	Oct. 20–22	30	$06^h 24^m$	$15°$	Fast meteors, broad peak.
Leonids	Nov. 15–20	Nov. 17	15	$10^h 08^m$	$22°$	Storms in 1799, 1833, 1866, 1966. Possible storms in 1998/1999.
Alpha Monocerotids	Nov. 15–25	Nov. 21	5	$07^h 48^m$	$01°$	Outbursts in 1925, 1935, 1985, 1995.
Geminids	Dec. 7–15	Dec. 13	120	$07^h 26^m$	$32°$	Finest of all the annual showers.
Ursids	Dec. 19–24	Dec. 22–23	10	$14^h 28^m$	$78°$	Outbursts in 1945, 1982, 1986.

is far from the inner Solar System. Although the comet returns to perihelion at intervals of 6.6 years, an orbital resonance means that favourable Giacobinid returns come only roughly every 12 years. The shower produced significant activity (50–450/h) on 1933 October 9 for a period of about 4.5 hours. Further activity was seen in 1946 and 1985. The 1998 return is considered somewhat unfavourable for Giacobinid activity, as Earth passes the descending node some 50 days ahead of the comet; only observations can give the true picture.

Unusual activity has occasionally been seen from other showers, though not always to storm levels. For instance, the Ursids around December 22–23 usually produce fairly modest rates, with a typical peak ZHR of about 15. In some years, however, the ZHR has exceeded 50. Notably, rates were high in 1946, and again in 1982 and 1986. Unlike those of the Leonids, Giacobinids and Perseids, the Ursids' outbursts do not seem to correlate with the perihelion of their parent comet, 8P/Tuttle. Peter Jenniskens, at NASA's Ames Research Center, has proposed a focusing mechanism involving the gravitational influences of Jupiter and Saturn to account for these events, which he describes as "far comet" outbursts.

A similar mechanism may account for the outburst seen from the normally fairly tame Lyrid shower on 1982 April 22, and for the 10-year periodicity of outbursts from the very minor Alpha Monocerotid shower. The Alpha Monocerotids normally produce activity barely distinguishable from the sporadic background. Outbursts were seen in 1925, 1935 and 1985, leading to several predictions of a recurrence in 1995. This was duly borne out on the night of 1995 November 21/22, when, for about 30 minutes around 01:30 UT, the shower produced rates of 2–4 meteors per minute. The display was recorded, uniquely, on videotape by Sirko Molau in Germany. Perhaps a similar mini-outburst awaits in 2005.

One of the most unexpected meteor outbursts in recent years came on 1998 June 27/28, as this book was in the final stages of preparation, from the Pons–Winneckid shower. Also known as the June Boötids, this shower is produced by debris from Comet 7P/Pons–Winnecke. Activity is usually indistinguishable from that of the sporadic background. Outbursts were seen in 1916, 1921 and 1926, after which perturbations of the stream's orbit dragged it away from the Earth. It therefore came as a surprise to observers to experience

Perseid-level rates from the shower for about 15 hours from 09:00 UT on June 27. Clearly, this shower will bear close scrutiny around this date in future years.

Undoubtedly, there will be further occasions on which otherwise well-behaved showers will produce unexpectedly high activity: glimpsing these is a reward for those dedicated individuals who make the effort to observe the activity of shower and background sporadic meteors whenever possible.

2.4 Fireballs

Most meteors seen visually are in the magnitude range +2 to +4, about the same as the magnitude distribution of easily detected naked-eye stars. In general, the proportions of brighter meteors are higher during the activity of the major annual showers: some, such as the Perseids and Geminids, are justly noted for their wealth of meteors at the bright end of the range, which makes them ideal targets for photography (see Figures 2.9 and 2.10).

Figure 2.9. A Perseid meteor recorded on 1983 August 13/14 at 01:34 UT. The track crossing the frame diagonally at the top right is of an artificial satellite.
Photograph by the author

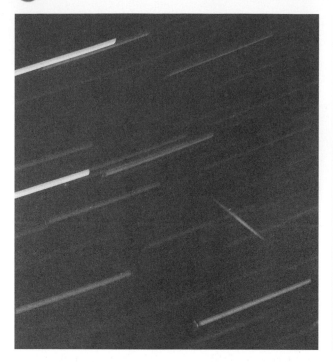

Figure 2.10.
A Geminid meteor captured on film on 1996 December 13/14 at 22:53 UT. *Photograph by the author*

The most spectacular meteor events of all are the rare fireballs, which may be seen at any time. A fireball is, by long-established convention, a meteor of mag. –5 or brighter – that is, brighter than the planet Venus. Fireballs account for perhaps 0.1% of all the meteors likely to be seen by a dedicated watcher. It follows that the chances of seeing such bright meteors are higher during the activity of the major showers, when rates are higher. Some of the most impressive fireballs, however, are sporadics, with no shower association.

The very brightest fireballs can have magnitudes comparable to that of the full Moon (mag. –12) or even brighter, for a few moments turning night into day. If they are slow-moving, such fireballs can last for several seconds and traverse a long arc across the sky. These events are often associated with the arrival of more substantial debris, usually of asteroidal rather than mereoroidal origin, which may sometimes survive atmospheric passage to fall to the ground as meteorites. Falls are rare, probably fewer than a dozen being witnessed each year, and each new example provides meteoriticists with potentially unique material for study. Fireball observations are therefore of considerable value: reports of the apparent paths of bright events, collected over reasonably widely separated locations, can allow an approximate real atmospheric trajectory to be determined, leading in turn to some idea of where any surviving material might have come to ground.

The most promising candidates for meteorite-dropping fireballs are slow, long-duration (several seconds in luminous flight) events. There is no recorded instance of a meteorite associated with a major meteor shower. In most instances, the meteoroids giving rise to shower activity are small "dust-balls" which are completely vaporised by frictional heating during atmospheric entry. Rocky debris, flung into Earth-crossing orbits following collisions and fragmentation in the main asteroid belt between Mars and Jupiter, is less friable and has a greater chance of surviving its passage through the atmosphere and reaching the Earth's surface.

With the exception of the Geminids and Taurids, each in short, direct orbits with aphelion no farther than the distance of the asteroid belt (2–3 AU from the Sun), most meteor streams have long elliptical orbits which reach out to or beyond the orbit of Jupiter. In keeping with Kepler's laws of planetary motion, we therefore tend to encounter meteoroids in a sector of the stream's orbit where, close to perihelion, they are travelling fastest, and their collisions with the upper atmosphere occur at very high geocentric velocities. The Perseids, Orionids and Leonids gain added collisional velocity since their orbits are retrograde, and therefore they impact the Earth "head-on". At high geocentric velocities (60–70 km s^{-1} collisions are not uncommon), the resulting meteors appear very swift and are of short duration.

Debris from collisions in the asteroid belt, on the other hand, may often have a lower collisional velocity. Some material, indeed, may even come in towards the Earth from behind in its orbit, at close to the theoretical minimum geocentric velocity of 11 km s^{-1}. Meteors or, more particularly, fireballs from this source will therefore appear slower-moving than most of their shower-derived counterparts. Additionally, the greater tensile strength of rocky material gives such fireballs their longer observed durations.

Over the area of the British Isles, perhaps one major fireball – an event of, say, mag. –8 or brighter – appears on average each month. In most cases, sadly, the event is seen by only one or two witnesses (often someone walking their dog in the evening), and little can be done other than to log the event for future reference. From the collected reports of many years it seems likely that some times of year are more productive of fireballs than others, excepting the activity of the major showers. February and March is one such "fireball season", while early December seems to be another. The early December peak in sightings may correlate with the very minor Chi Orionid meteor shower, which typically produces rates of only a couple of meteors per hour. The spring peak may be the result of a stream of loosely related asteroidal debris following similar orbits and left over from a collision a long time ago. Several *asteroid families* resulting from collisions have been identified among the main-belt population, and it is not unlikely that similar families of much smaller debris, in Earth-crossing orbits, also exist.

The fireballs that cause the most interest and excitement, and are most rewarding to follow up, are those seen by large numbers of witnesses over a wide area. In such cases it is important to collect, as soon as possible after the event, eyewitness reports, from which it may prove possible to reconstruct the fireball's trajectory. The relevant details include time of appearance and apparent brightness, and the altitude and azimuth of the fireball's path across the sky.

Magnitude estimates for brilliant fireballs can only be approximate. As a guide for comparison, Venus is of mag. −4, the first quarter Moon mag. −8, full Moon mag. −12, and the Sun mag. −27. Where events are particularly bright there is little more than academic interest in the magnitude in any case!

Observers will often describe the shape and apparent size of fireballs, and these reports can be useful. Many fireballs appear as teardrops, perhaps trailing "blobs" of detached material behind the head. Fragmentation in flight does not bode well for the recovery of meteoritic material. Sometimes, at the end of its luminous flight, a fireball will explode as the incoming meteorite shatters under aerodynamic stress in the low stratosphere at altitudes of 15–30 km. These events can lead to smoke trails in the sky which may persist for some time. One example of such an event was a daylight fireball seen over southeast England and from the north of Holland on the morning of 1994 May 29. The terminal explosion may be followed, after a few minutes, by a low rumbling as sound waves from the detonation reach the observer.

Probably the most useful elements of any fireball report are estimates of the altitude and azimuth, in degrees, of the event in the observer's sky. Azimuth is reckoned in degrees from due north (000°), through due east (090°), and so forth. Altitude is measured from the horizon (0°) to the zenith (90°). If estimates of the apparent start and end of a fireball's path in these coordinates can be obtained from several widely geographically separated observers, it becomes possible to triangulate its real trajectory. Ideally, observations should be available from either side of the track.

In several countries, photographic patrols for fireballs, using wide-angle fisheye lenses to maximise sky coverage, have been carried out with varying degrees of success. Among the more productive camera networks is that operated in central Europe. The American Prairie Network, operated by volunteer farmers during the 1960s, scored a notable success with the recording of the fireball associated with the Lost City Meteorite, whose trajectory was calculated from multiple-station records, leading to its recovery. Weather conditions play an important part in determining the success of a fireball patrol network. Attempts to run a full network over the British Isles, for example, seem always to have been frustrated by clouds affecting some part or other of its range. However, the capture of many excellent images of fireballs each year, including the spectra of several, by Henry Soper on the Isle of Man in the often cloudy region of the Irish Sea should stand as ample encouragement to those prepared to make the effort!

References and Resources

Bone, N, *Meteors*. George Philip/Sky Publishing (1993).

Bone, N, "Meteors" in *The Observational Amateur Astronomer,* edited by P Moore. Springer (1995).

Bone, NM and Evans, SJ, "Visual and photographic observations of the Perseid meteor shower in 1993". *Journal of the British Astronomical Association* 106 1 33–9 (1996).

Bone, NM and Evans, S, "Visual and photographic observations of the Perseid meteor shower in 1994". *Journal of the British Astronomical Association* 107 3 131–5 (1997).

Leonid 98 Meteor Outburst Mission Homepage: http://www-space.arc.nasa/~leonid/

Mason, JW, "The Leonid meteors and comet 55P/Tempel–Tuttle". *Journal of the British Astronomical Association* 105 5 219–35 (1995).

Rao, J, "The Leonids: King of the meteor showers". *Sky & Telescope* 90 5 24–31 (1995).

Chapter 3
Aurorae and other Atmospheric Phenomena

Regular observers of the night sky are among those most familiar with the various short-lived cloud and optical phenomena to be seen in the Earth's atmosphere. Amateur astronomers keeping a weather eye on the sky before nightfall to gauge whether a clear night is likely are often in a position to record atmospheric halos or parhelia ("sundogs") produced by the refraction of sunlight by ice crystals in advancing veils of cirrus cloud, for example, or unusual formations such as lenticular clouds (Greenler, 1989; Minnaert, 1954). Many amateur observers marvelled at the succession of purple twilights in 1991 and 1992 which resulted from the injection of volcanic material into the stratosphere by the eruption of Mount Pinatubo.

On rare occasions the avid skywatcher may be treated to more exotic displays. For instance, observers in the British Isles recorded a rare apparition of *nacreous clouds* on the afternoon and evening of 1996 February 16. These clouds, named for their mother-of-pearl iridescence, form in the stratosphere at altitudes of about 30 km, and are rarely seen outside the polar regions. For all their visual beauty, nacreous clouds are believed to be sites where ozone-depleting chemical reactions take place, leading, in the Antarctic, to the infamous ozone hole.

Cloud phenomena, however, lie strictly in the realm of meteorology rather than astronomy. Other phenomena, occurring at greater altitudes in the atmosphere, do have astronomical connections and deserve our attention.

3.1 Auroral Storms

Among the most awe-inspiring of short-lived astronomical phenomena are the major auroral displays which, several times each sunspot cycle, may penetrate to lower latitudes from their more customary polar habitats. Displays during the 1980s and 1990s have demonstrated to amateur astronomers in the southern United States and the more southerly parts of the British Isles that the aurora may occasionally be witnessed from such normally unfavourable locations (Table 3.1). The huge auroral storm of 1989 March 13/14 was widely seen around the world, as were later events on 1991 March 24/25 (Figure 3.1) and November 8/9 (Figure 3.2). These each had their roots in violent activity in the Sun's inner atmosphere.

As shown in Figure 3.3 (page 36), auroral activity at lower latitudes roughly follows the solar cycle, being commonest a year or so ahead of sunspot maximum, with a second peak following about a year after maximum. Sunspot maximum itself appears to be a quiet time for aurorae at lower latitudes. On this basis, we might expect 1998–99 and 2001–2002 to be productive times for aurorae at lower latitudes, as sunspot activity in Solar Cycle 23 took off in mid-1997.

Figure 3.1. The aurora of 1991 March 24/25 as seen from Chichester in southern England. Strong red rays filled the north-eastern sky. *Photograph by the author*

Figure 3.2. Rayed band aurora on 1991 November 8/9, photographed from Chichester. *Photograph by the author*

Globally, auroral activity is almost always present in two oval regions, one surrounding each geomagnetic pole. Observers at high latitudes, close to the auroral ovals, can see activity on a more or less nightly basis. The northernmost land masses are favourably distributed for observing the northern auroral oval, whose behaviour can be monitored continuously from locations such as northern Canada, Alaska and the North Cape of Scandinavia. Owing to their higher geomagnetic latitude (resulting from the

Table 3.1. Major aurorae of Solar Cycle 22 (1986–96)

Date	Lowest Latitude of Visibility
1989 March 13/14	Grand Cayman
1989 October 20/21	Southern England, southern US
1989 November 17/18	Southern England
1990 July 28/29	Southern England
1991 March 23/24, 24/25	Southern US, northern France
1991 June 4/5	Central US
1991 June 10/11	Central US, southern England
1991 October 1/2	Northern Italy
1991 November 8/9	Southern France, deep south US
1991 December 29/30	Southern England
1992 February 2/3	Southern US
1992 July 24/25	Southern England
1992 August 22/23	Central US

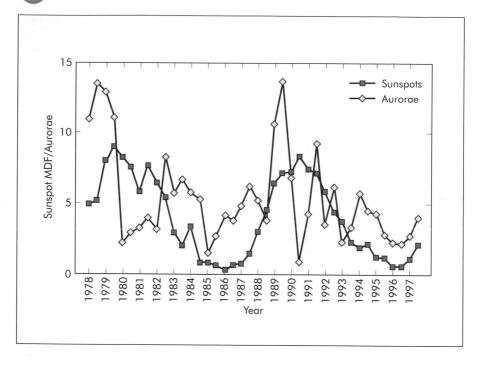

offset between the geomagnetic and rotational axes), American observers are better placed than their European counterparts.

3.1.1 The Causes of Auroral Activity

The brightness, extent and activity of the auroral ovals depend on conditions in the magnetosphere – the volume of space surrounding the Earth in which its magnetic field exerts a dominant influence on the movement of charged particles. The Earth's magnetic field, generated by electrical currents in the planet's molten metallic core, presents a barrier to the solar wind – the stream of ionised particles (mostly protons and electrons) that flows outwards from the Sun and past the planets. The solar wind compresses the magnetosphere on the sunward side, and draws it out downwind into a comet-shaped structure called the magnetotail. Events in the magnetotail are believed to govern the state of activity in the auroral ovals.

The solar wind varies in intensity, and (most importantly, in current models of auroral generation (Akasofu,

Figure 3.3. Auroral activity as a function of sunspot activity. Average monthly numbers of aurorae visible in Britain south of the Orkneys for six-month intervals are plotted together with average sunspot mean daily frequency (MDF) for the same period. In general, low-latitude auroral activity follows the sunspot cycle, though there is – paradoxically – a marked dip at sunspot maximum. Aurorae at lower latitudes are commonest a year or so ahead of, and about 18 months following, sunspot maximum.

1989)) in the orientation of the entrained magnetic field which it carries along with it. Variations in this magnetic field, determined by features at or close to the Sun's surface, can in turn stress the magnetosphere by influencing the efficiency of coupling between the solar and terrestrial magnetic fields. On average, the solar wind flows past the Earth at a velocity of 400 km s^{-1}.

Violent eruptions in the inner atmosphere of the Sun can create "gusts" in the solar wind of up to 1000 km s^{-1} which buffet the magnetosphere on their arrival, perhaps a day or so later, in near-Earth space. Major variations in the solar wind arise from solar flare activity and from coronal mass ejections (CMEs). The relationship between these phenomena has been the subject of much recent professional study, leading to the proposal that CMEs may cause solar flares rather than vice versa (Crooker, 1994). The highly successful SOHO satellite, stationed at the stable L$_1$ orbital point 1.5 million km upwind of Earth, has imaged a great many CMEs, even during times of relative solar quiescence. It is a long-term aim of professional scientists to be able to give early warning of potential geomagnetic and near-Earth disturbances following CMEs, and progress is being made towards providing such "interplanetary weather forecasts".

Whatever their cause-and-effect relationship with solar flares, CMEs lead to the injection into the solar wind of pockets of increased velocity, particle density and magnetic field intensity. Enhanced coupling between the magnetosphere and the solar wind at these times increases the flux of particles entering near-Earth space, placing enormous stress on the magnetotail. Under such stressed conditions, electrons from the magnetotail's plasma sheet are accelerated into the high atmosphere, brightening the auroral ovals on their night side. Disturbances of the Earth's magnetic field under these conditions may also be severe. The effects of these major geomagnetic storms, resulting from magnetospheric currents, can be detected by magnetometers, and may even prove to be of commercial importance. Damage to satellites has been reported following some geomagnetic storms, while the 1989 March 13/14 event was accompanied by ground currents which caused power outages in Quebec Province, Canada (Schaefer, 1997).

For the visual or photographic observer the main appeal of the aurora is, of course, the shifting pattern of often vividly coloured light in the poleward sky. The light emission has its origins in the laws of quantum physics. As accelerated electrons rain into the tenuous upper atmosphere during disturbed geomagnetic conditions, atoms and molecules of oxygen and nitrogen become stripped of electrons in the permitted energy levels around their nuclei. Electron capture to refill these quantum shells leads to the emission of light at very specific wavelengths. Auroral light is dominated by oxygen emission at wavelengths of 557.7 nm in the green part of the visible spectrum and 630.0 nm in the red, and purplish nitrogen emission at 391.4 and 427.8 nm. All these emissions can occur only under conditions of low gas density: typically, auroral emissions originate at altitudes in excess of 100 km.

3.1.2 Auroral Substorms

Observers at high latitudes, close to the auroral ovals, are familiar with substorm activity, which may occur several times a night even under quiet solar conditions. This activity, observable as a brightening of the night side of the auroral oval accompanied by a westward-travelling surge which carries arcs and bands along with it, is thought to originate in comparatively minor fluctuations in the solar wind. These fluctuations put much less stress on the magnetotail than do those that follow major CMEs. Under normal conditions there may be between three and five substorms a day.

As observed from the latitudes of Alaska or Northern Canada, substorms follow a fairly regular pattern. A quiet evening arc of aurora may suddenly brighten, or become multiple, followed by an outburst of strong rayed activity with folded, rippling bands giving the appearance of the oft-portrayed auroral "curtains" moving swiftly across the sky. Rapid changes in brightness – flickering or flaming – and streaming of rays along the arcs or bands may mark the height of the disturbance, lasting an hour or two before the display falls back and the aurora becomes quiescent again (Davis, 1992). Some observers have found it profitable to travel to higher latitudes to observe substorm activity, the equinoxes being a particularly good time. Obviously, the aurora is unlikely to be seen in the twilit summer months at high latitudes, while such locations may also be prone to spells of severe, observationally unfavourable weather.

3.1.3 Geomagnetic Storms

While substorm activity is, by and large, a dependable, regular feature of the dark high-latitude nights between autumn and spring, the great geomagnetic storms that bring the aurora to lower latitudes are (as yet) far from predictable. Certainly, the big storms are generally transient events: if major aurora has been seen in the southern United States or England on one night, it is unlikely to be present the next. The would-be observer needs either to be patient, and carefully watch the sky for activity (most major aurorae show themselves, unexpectedly, to those who are outdoors observing other things), or to rely on early warning via their local astronomical society or a more extensive "ring-round" alert network. Several such networks have sprung up since the great aurora of 1989 March. A couple of Web sites offer auroral forecasts, but auroral prediction still remains a black art. Obviously, high solar activity is a useful indicator; flare activity at hydrogen-alpha wavelengths (Section 4.2), seen close to the central meridian of the visible solar disk, may precede auroral activity by a day or so.

During major geomagnetic storms vast amounts of energy are transferred into the ionosphere at auroral heights. The auroral ovals brighten, and they expand markedly, especially on the night side, towards the equator. Under extreme conditions the northern-hemisphere oval may be extended to the Mediterranean or beyond, as happened in 1989 March. At such times observers at lower latitudes have a brief opportunity to witness the awesome spectacle of the aurora far from its "normal" locale.

While it is extremely rare for the auroral ovals to be pushed to such extremely low latitudes as they were during the 1989 March event, geomagnetic disturbances sufficient to carry the aurora to visibility from Scotland or the northern United States arise reasonably frequently in the years around sunspot maximum. There is no guarantee, however, that a trip to such locations, made with the specific intention of observing the aurora, will bear fruit.

Auroral storms at higher mid-latitudes often follow a fairly consistent activity pattern. The display may begin in the evening as a glow – comparable to the pre-dawn twilight, hence the term *aurora borealis*, "northern dawn" – over the poleward horizon. This may, indeed, be the total extent of the display if the disturbance fails to develop much further; observers closer to the auroral oval will have a more impressive view. Quite often, however, the glow rises higher into the sky, and brightens. It may go on to develop into a more discrete *arc* structure, spanning east–west across the sky, with its highest point more or less in the direction of the pole.

Episodes of fading or brightening may follow, leading to the development of vertical *ray* structures. The arc will often fold into a ribbon-like *band* at this stage. Rays frequently move along the length of arcs or bands, giving the appearance of curtains rippling across the sky.

At times of very high activity the aurora may fill the whole of the poleward sky, or even stretch to the zenith or beyond. On those occasions, rare at mid-latitudes, when the aurora comes to fill the whole sky, the effect of perspective comes into play and the essentially parallel rays may appear to converge on a single part of the sky, taking on the form of a *corona*.

Corona formation marks the peak of a disturbance; if seen at all, it may persist for only a few minutes before the display falls back polewards. In most displays seen from central Scotland or the mid-latitudes of the United States, the display's peak may be the rayed band stage, not reaching as far as the zenith.

In major storms, activity sometimes shows repeated peaks, and the observer should not assume that the decline of a rayed outburst marks the end of a display. After an interval of an hour or so, the aurora may build in intensity once more. Major storms often show several peaks of activity in the course of a night, each building from a quieter interlude.

3.1.4 Observing the Aurora

When it occurs at lower latitudes, the aurora may occupy a considerable area of sky. The best way to view it is undoubtedly with the naked eye; there is no point whatsoever in trying to observe the basically diffuse light of auroral features through any sort of optical instrument.

Visual observers are encouraged to report sightings of the aurora to the appropriate section of their national amateur body. Many in north-west Europe, for example, submit reports to the Aurora Section of the BAA. Reports may consist simply of the information that aurora was visible from a given location on a particular night. Statistics of auroral visibility as a function of time in the solar cycle have been built up over a great many years, and such reports add to this larger picture.

Experienced observers may prefer to give a more detailed account of the varying activity of a display as the night progresses. A detailed report will give an indication of the particular auroral forms present, and the time (in UT) when they were seen. The extent of features in azimuth and altitude should be reported. Thus, a quiet early-evening arc may span from 300° to 060° azimuth, reaching a maximum altitude above the horizon (conventionally denoted by ↗) of 15°. Another useful measurement (denoted by h) indicates the altitude above the horizon of the highest point on the base of a feature; so, for instance, the arc mentioned above may have $h = 10°$. Since auroral emissions cut off quite sharply around an altitude of 100 km, measurements of h allow a reasonable estimate of geographical extent to be made.

Some observers produce highly detailed blow-by-blow accounts of aurorae as they appear from one minute to the next. In practice, however, it is usually sufficient to note the display's extent and appearance only as and when it changes. Changes in brightness, which range from slow pulsations to more rapid vertical "flaming" or lateral "flickering", should also be noted. Auroral brightness is indicated on a four-point scale from I (very faint), through II (comparable to moonlit cirrus), III (moonlit cumulus) and IV (very bright, perhaps even casting shadows!). Colour may often be seen, particularly during very big displays. Greenish emissions usually predominate lower in the display, with pronounced reds near the tops of rays high in the atmosphere.

Naturally, the aurora is a superb subject for photography. While a good record can be obtained on fast black-and-white film, it is much more satisfying to photograph the aurora in colour. Favoured emulsions with auroral photographers are Kodak Ektachrome 400 and Fujichrome 400, both fairly fast colour slide films. These give a reasonably faithful rendition of auroral colours; other films have been found to over- or underemphasise auroral greens and reds, depending on their spectral sensitivities.

The camera, with either a standard 50-mm lens or – many observers' preference – a wide-angle 28-mm lens, open as wide as possible (ideally $f/2.8$ or less), should be firmly mounted on a tripod for time exposures, controlled by a cable release. Bright, rapidly moving aurorae can be captured in exposures of 10–15 seconds' duration or even less. Fainter, more static glows or arcs may require exposures of 30 seconds or more to record them well. There are no rigid rules in auroral photography. The best advice one can offer to low-latitude observers in particular is that opportunities to experiment will be few and far between, so do not stint on film when a display is present.

3.2 Noctilucent Clouds

While clouds in the lower atmosphere are the bane of the amateur astronomer's life, many observers in north-west Europe, Canada and the more northerly parts of the United States actively seek out one form of high-atmosphere cloud during the summer months. These *noctilucent clouds* – literally, "night-shining" clouds – form at very much greater heights than the feathery cirrus to which they bear some resemblance, and are probably the result of trace quantities of water vapour condensing onto small particles of meteoric debris close to the mesopause, at altitudes of 82–85 km. Noctilucent clouds (commonly abbreviated to NLC) are commonest in the years around sunspot minimum, presumably as a result of the generally lower atmospheric temperatures prevailing at these times. Ultraviolet and X-ray emissions from active regions around sunspot maximum heat the upper atmosphere, making conditions less favourable for NLC condensation. NLC are a summertime phenomenon, appearing at a time of year when the upper atmospheric circulation carries water vapour to their sites of formation around the polar cap, and when temperatures there reach their annual minimum.

NLC are extremely tenuous and cannot be seen by daylight. The thin, delicate sheets become visible late on summer evenings when the Sun is between 6° and 16° below the observer's horizon. At such times the lower atmosphere is immersed in the Earth's shadow, while NLC remain in sunlight, appearing bright in contrast with the twilit sky (Figure 3.4). Favourable illumination conditions persist for much of the brief summer night at the latitudes of Scotland or Canada in the NLC

Figure 3.4. A strong display of noctilucent cloud over Sussex on 1993 June 28/29. *Photograph by the author*

season of June and July. The season seems to have a peak around the solstice on or around June 21, most displays being reported between the third week of June and the second week of July. From north-west Europe NLC have been seen as early as mid-May and as late as mid-August.

NLC displays are distinctive. The clouds have a silvery-blue colour unlike any others, and often show a strong banded structure which may become obvious on inspection with binoculars. Clouds in the lower atmosphere appear dark in contrast against NLC displays.

Displays quite often persist throughout the short summer night, and from the most favoured latitudes (around 57°N) may be seen on several successive nights around the peak of the season. At lower latitudes displays are less frequent; observers in the northern United States or southern England might be lucky to see a couple of NLC in a summer. Few reliable sightings are made south of 50°N latitude.

3.2.1 Observing Noctilucent Clouds

Reports of NLC are collected by the BAA Aurora Section in north-west Europe, and in North America by the NLC Can-Am Network. Unlike auroral displays, NLC tend not to change rapidly. The main influence on their appearance is the gradually changing solar illumination, displays generally being most extensive in early evening and just before dawn, and sinking towards the northern horizon as the Sun nears its lower transit at midnight. It is therefore normally sufficient to record a display's appearance at 15-minute intervals.

Reports should indicate the extent in altitude and azimuth, exactly as for auroral features (Section 3.1.4). Like the aurora, NLC show a range of fairly distinctive forms, and it is useful to note which are present:

Type I   *Veil*: very tenuous, lacking in structure, often a background to other forms.

Type II   *Bands*: long streaks, often in groups, parallel or crossing at small angles.

Type III *Billows*: closely spaced, resembling waves or ripples, very characteristic herring-bone structure.

Type IV   *Whirls*: large-scale looped structure, often as complete or partial rings.

Type V   *Amorphous*: similar to veils in their lack of structure, but brighter, usually in patches.

Displays often take the form of parallel bands (Type II). NLC displays frequently show a highly organised wave pattern which, together with their colour, sets them apart from cirrus and helps to identify them. It can be useful to add rough, annotated sketches of a display's appearance to the report.

From the more southerly parts of their region of visibility, NLC seldom reach elevations greater than 10°, about the same as the bright star Capella which skirts low across the northern sky on summer evenings. From more northerly latitudes, however, displays can be very extensive, occasionally covering the entire sky.

NLC are obvious targets for photography and, as with the aurora, results on colour film are most attractive. Using a standard 50-mm lens at $f/2.8$, exposures of 1–3 seconds on ISO 400 film should suffice to record the cloud-forms. Bright displays will require shorter exposures. It is advisable to keep exposures reasonably short to avoid the cloud details becoming swamped by the twilight. Kodak Ektachrome 400 film has given excellent results over the years, and often brings out the clouds' silvery-blue tints, shading off to a sunset gold as a result of atmospheric reddening close to the horizon. Since NLC often appear quite close to the horizon, the photographer is more or less obliged to include some of the foreground, adding to the visual appeal of the images.

NLC observers in the British Isles have long been encouraged to take exposures at standard times: *precisely* (to the second) on the hour, half-hour and quarter-hours. Photographs of NLC displays obtained in this way from reasonably well-separated locations can be used to triangulate cloud features, allowing their altitude, extent and movements to be determined.

Much remains to be learned about the processes in, and circulation of, the atmosphere at the great heights at which noctilucent clouds occur. Amateur observations remain of value to professional atmospheric scientists, and the pursuit of these clouds is a useful activity for those at higher latitudes whose other astronomical activities are restricted by the prevailing twilight on summer nights.

3.3 Synthetic Atmospheric Phenomena

Observers in Finland and Estonia have occasionally reported unusual, iridescent NLC-like displays following Russian rocket launches during the summer months. These appear to result from the exhaust emissions. A similar "artificial noctilucent cloud" followed the launch, from Cape Canaveral in Florida, of the Cassini probe to Saturn on 1997 October 15 (McGee, 1997). Exhausts from rocket launches either seeding, or condensing as, clouds in the high atmosphere may account for several reports of NLC at anomalously low latitudes.

Satellites in low orbit are often visible to the naked eye as points of light moving across the sky. Viewing conditions are particularly favourable when the Sun is not far below the observer's horizon, and many satellites are visible in the early-evening twilight or just before dawn. During the summer months at higher temperate latitudes, the same illumination conditions which favour the visibility of NLC mean that a large number of satellites may be seen; scarcely a few minutes pass without a satellite becoming visible in some part of the summer sky.

A host of objects have been placed in orbit around the Earth since the dawn of the Space Age in 1957, ranging from small satellite or rocket nose cones, through fragments from disintegrated booster stages, and substantial bodies like the Mir space station. Mir can appear as a very bright "star", of magnitude –2 or brighter under some illumination conditions. From time to time, formations – triangles or parallelograms in close orbital proximity – of satellites flown together for geophysical research may be seen by the alert observer. Meteor observers in Europe became familiar with a triangular flight of satellites known as the NOSS trio, each of visual mag. +3, during the summer of 1997.

On time-exposure photographs, satellites of about mag. +3 or brighter are fairly readily recorded on ISO 400 film with a standard 50-mm or 28-mm wide-angle lens operating at $f/2.8$ or faster. The resulting trails relative to the star background are a source of considerable frustration to meteor photographers (see Figure 2.9). Indeed, many alleged "meteors" on exposures of Comet Hale–Bopp obtained on the twilit spring evenings of 1997 are, in reality, trails of satellites in sunlight at orbital altitudes in the western sky.

Many satellites rotate, leading to changes in their apparent brightness as reflective surfaces or solar panels are brought into differing alignment with the Sun. Slowly rotating satellites may sometimes appear to flare in brightness over a second or two (appearing almost like point meteors). On photographs, such objects can appear quite convincingly like meteors. Flares from members of the "constellation" of Iridium communications satellites (Chien, 1998) can reach mag. –7, almost as bright as the half Moon, and are regarded – with some justification – as a hazard to sensitive optical equipment such as photomultipliers.

While meteor observers and those photographing the deep sky with the aim of obtaining attractive pictures regard satellites as little more than a nuisance, there is a small hard core of observers who record the passage of satellites against the star background for scientific purposes. Precise timings of points along a satellite's traverse across the sky can be used to determine how its orbit is affected in the medium term by the Earth's gravitational field and tenuous upper atmosphere. Atmospheric drag will eventually, unless counteracted by firing rocket motors, lead to a satellite's decay (re-entry) from low orbit. The resulting fireballs may sometimes be spectacular, and often last longer than those associated with the arrival of meteoritic debris (Section 2.4).

The firing of attitude motors on artificial satellites has, on occasion, been the source of alerts of unusual transient phenomena. For example, on 1981 March 14 Kelemen Janos and other observers in Hungary, and J. Bremseth in Norway, observed what they initially suspected to be a "mini-comet" passing rapidly through near-Earth space. From Hungary, the object appeared as a 0.5°-diameter circular patch of mag. +3.5 to +4.0, with a central condensation or "nucleus". Seen from Norway, the object was brighter, about mag. +1 to +2. Motion was quite rapid, and subsequent investigation led to the identification of the object as an experimental Russian hunter-killer satellite, Cosmos 1258, sent to target a previous launch, Cosmos 1241. The glow surrounding Cosmos 1258 was caused by the photoionisation by sunlight of exhaust gases from its attitude motors.

Satellites in geostationary orbit have for many years been a mainstay of global communications and broadcasting. The delivery of satellites to these distant (35 900-km altitude) orbits requires the use of booster rockets to carry them from low orbit. Firings of the apogee motors on rockets used to raise satellites to high orbits have also led to some unusual sightings. Among the best-documented was another suspected near-Earth "comet", seen from Germany and eastern Europe on the night of 1994 May 3. This object appeared as a brilliant (mag. –4, as bright as Venus) V-shaped glow very similar in shape to a comet, among the stars of Perseus, fading slowly over the course of an hour or so. The source of this sighting was the firing of an apogee motor aboard a US military Centaur rocket, carrying a secret military satellite launched from Cape Canaveral earlier in the day.

Researchers investigating processes and particle movements in the ionosphere between altitudes of 140 and 400 km occasionally carry out experimental releases from sounding rockets. On release at high altitudes, barium vapour rapidly becomes photoionised by solar ultraviolet radiation, fluorescing green and violet. Barium cloud releases allow the near-Earth magnetic field to be traced out. Visually, the clouds are dominated by the green emission; the violet is difficult to see. The experiments are usually carried out during the summer months when the high atmosphere is in sunlight and it is dark at ground level – the same illumination conditions that favour NLC and satellite observations.

Barium cloud releases are often carried out over Alaska, using rockets launched from the range at Poker

Flats. Similar experiments were performed over Scotland in the early 1970s, with launches from the Hebridean island of South Uist. These releases were seen by ground observers as elliptical, slowly expanding greenish clouds.

Investigations of the solar wind in near-Earth space have been carried out using gas releases from satellites. These, too, can sometimes be visible to the ground-based observer. In the international AMPTE project in 1984, for example, releases were made outside the Earth's magnetosphere to create an artificial "comet" in the solar wind. The CRRES satellite launched in 1990 performed a series of gas releases which were observed by amateur astronomers in the United States during January 1991.

References and Resources

Akasofu, S-I, "The Dynamic Aurora". *Scientific American* 260 5 54–63 (1989).

Astronomy 19 6 30 (1991) [photographs of CRRES experimental gas releases].

Aurora Alert Hotline: David Huestis, 25 Manley Drive, Pascoag, RI 02859, USA. Tel. (401) 568–9370.

Aurora Web Site: http://www.geo.mtu.edu/weather/aurora

Bone, N, *The Aurora: Sun–Earth Interactions*, 2nd edition. Wiley-Praxis (1996).

Bone, N, "Seeing the summer twilight clouds". *Astronomy Now* 9 7 22–4 (1995).

Chien, P, "Have you been flashed by iridium?" *Sky & Telescope* 95 5 36–41 (1998).

Crooker, N, "Replacing the solar flare myth". *Nature* 367 595–6 (1994).

Davis, N, *The Aurora Watcher's Handbook*. University of Alaska Press (1992).

Gadsden, M, "Can I see noctilucent clouds?" *Journal of the British Astronomical Association* 108 1 35–8 (1998).

Greenler, R, *Rainbows, Halos and Glories*. Cambridge University Press (1989).

McGee, H, "Cassini launch is 'picture-perfect'". *Journal of the British Astronomical Association* 107 6 338–9 (1997).

Minnaert, M, *The Nature of Light and Color in the Open Air*. Dover (1954).

NLC Can-Am: Mark Zalcik, 9022–132 A Ave, Edmonton, Alberta, Canada T5E 1B3.

NLC Web Site: http://www.personal.u-net.com/~kersland/nlc/nlchome/htm

Schaefer, BE, "Sunspots that changed the world". *Sky & Telescope* 93 4 34–8 (1997).

Chapter 4

The Sun

As the primary body of our Solar System, containing about a thousand times the mass of the planets and all the other bodies in orbit around it, our Sun is of obvious, fundamental importance to the Earth and the life upon it. The Sun's heat drives our global weather patterns, while its light is utilised by green plants for the essential process of photosynthesis. Activity at, or close to, the surface of the Sun, as we have already seen (Section 3.1.1), can have a major influence on the solar wind flowing past the Earth, in turn producing geomagnetic storms. Such events can affect communications and other systems, and will take on greater significance when the near-Earth orbital environment comes to be used more extensively for human activities – most particularly the operation of the permanently manned International Space Station. Energetic particles arriving in this environment following solar flares or coronal mass ejections present a hazard to orbital hardware and to living tissue above the atmosphere's protective shield. Given the importance of such "space weather" to the future exploitation of the orbital environment, it is hardly surprising that professional scientists give a high priority to the continuous monitoring of solar activity in an attempt to improve their understanding and forecasting of it.

Such round-the-clock study of the Sun, using large dedicated telescopes and spacecraft such as SOHO, leaves little original research to be done by amateurs using small instruments. Nevertheless, there is much of interest to be seen and considerable personal satisfaction to be derived from being aware of the current state of solar activity, which varies on a number of time-scales. If nothing else, observing the Sun is one obvious astronomical activity which does not have to cease during the daytime!

4.1 Observing the Sun in White Light

It goes without saying that the Sun is large. Its actual diameter of 1.4 million km translates to an apparent disk in Earth's sky of angular diameter between 31′ 31″ when we are aphelion in early July, and 32′ 35″ at perihelion in early January.

The Sun is an abundant source of light and heat radiation. As such, it needs to be treated with respect by the observer. Magnified through any optical system, the Sun's light will instantly cause irreparable damage to the eye. The observer should *never* look at the Sun directly through a telescope or pair of binoculars. Eyepiece filters should certainly not be trusted – they may crack in the heat of the concentrated sunlight near the telescope's focus. The safest way to observe the Sun in white light is to project its image onto a piece of clean, white card, as shown in Figure 4.1.

Figure 4.1.
Projecting the Sun's image onto card is the safest way to observe sunspots and other features such as faculae, or – as shown here by John Palmer (left) and Ian Merritt of the South Downs Astronomical Society – a partial solar eclipse. *Photograph by the author*

A small telescope is sufficient to give a good, clear projected image of the Sun. Indeed, there is little advantage in using a larger instrument, which will generally need to have its aperture stopped down to reduce potential heat damage to the eyepiece. A projected disk diameter of 150 mm is used as a standard by most amateur observing groups. With a very small telescope better results might be obtained by using a 100-mm disk. The projected image will reveal the presence of two principal features – *sunspots* and *faculae* – depending on the stage of the solar cycle. These features are both related to magnetic activity on the Sun.

Sunspots appear as dark regions against the bright solar surface, the *photosphere*. The photosphere has a temperature of around 6000 K, and at times when the air is very steady (most often when observations are made soon after dawn, before the Sun's heat has had time to stir up atmospheric turbulence) it may appear to have a fine mottled pattern, or *granulation*. The granulation results from small-scale convection in the outer layer of the Sun. Sunspots appear dark against the photosphere by contrast, thanks to their lower temperatures, around 4000 K, and form in regions where concentrations of magnetic flux emerge from the solar interior, disrupting the local convective pattern in such a way that gas on the surface cools. Sunspots are, of course, cool only in comparison with their surroundings. Seen in isolation against a night sky, a typical spot would shine with a reddish light, brighter than the full Moon.

Sunspots range in size from small *pores*, perhaps 1000 km in diameter, to massive, complex groups which may cover a substantial area many times the size of the Earth. The larger spots usually show structure, having a dark, central *umbra* surrounded by a lighter, greyish region of *penumbra*. The penumbra often has the appearance of radial filaments. Large, complex groups arise most often in the years close to sunspot maximum.

The sunspot cycle, in which the number of spots varies over a roughly 11-year period, has been known since the mid-19th century. Cycles are numbered in a series from the time when sunspot numbers can first be assessed sufficiently reliably from historical records. Cycle 21 of the series peaked in 1980, and Cycle 22 in 1989–90; Cycle 23 got under way in 1996–97 and is expected to peak around 1999–2000. In general, the rise of the sunspot cycle to its maximum is more rapid than the subsequent decline (Figure 4.2, *overleaf*). Also,

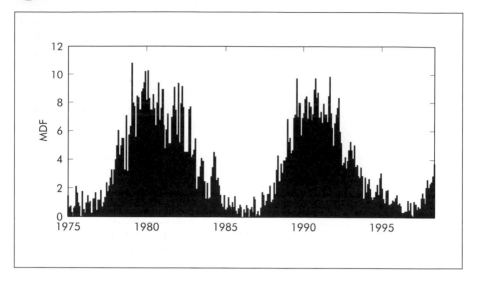

Figure 4.2.
Sunspot mean daily
frequencies (MDFs)
from January 1975 to
March 1998. The
maxima of Cycle 21 in
1979 and Cycle 22 in
1990 are clearly seen,
along with the minima
of 1976, 1986 and
1996.

maxima vary in intensity: Cycle 19, for example, showed a very strong peak indeed in 1957–58, while the following maximum in 1969–70 was comparatively weak. Evidence can be found in Chinese and other pre-telescopic annals for the long-term operation of the sunspot cycle, but there is also evidence of intervals during which spots were absent altogether – notably the so-called Maunder Minimum of 1645–1715.

Sunspots are the most visible manifestation of activity on the Sun, stemming from an overall magnetic cycle which takes 22 years to return to its starting configuration. The mechanisms by which sunspots arise remain to be fully understood, but they appear to involve interactions between the solar magnetic field and the differential rotation of the Sun. At the equator, the Sun takes about 25 days to rotate on its axis, whereas the period at the poles is closer to 36 days. This differential rotation is thought to cause the Sun's magnetic field to become wound around itself in the plane of the equator, resulting in magnetic flux "tubes" being forced to emerge through the photosphere in the regions where sunspots form. In a given cycle, sunspots initially break out at comparatively high solar latitudes (about 45°N and 45°S). As the cycle advances, spot groups emerge progressively closer to the equator. There is some degree of overlap between cycles, so that spots from the old cycle may be seen near the equator at the same time as those of the new cycle are beginning to break out at high latitudes.

The most favoured method of amateur solar observation is to make drawings of the projected disk, on which the sunspot positions are accurately recorded. Details of surrounding penumbra, and of individual umbral centres in complex groups, are added as the drawing progresses. Drawings should be made in this way each day when possible, and the time (in UT) of the observation recorded.

Such drawings can be used in conjunction with standard graduated disks, taking account of the Sun's varying apparent axial tilt over the course of the year, to allow heliographic positions to be obtained for sunspots. Probably the most widely used are the eight Stonyhurst disks. In order to use these, the observer needs to know the heliographic latitude of the centre of the visible solar disk, denoted by B_0, and listed in annual handbooks. B_0 reaches a maximum of +7.3° when the north pole of the Sun is tilted towards us around September 9, and –7.2° when the south pole is tilted towards us around March 7. B_0 is zero around December 7 and June 6. In addition to B_0, the observer also needs to take account of P, the position angle of the Sun's north pole relative to the ecliptic vertical, which may be as much as –26.3° in early April (apparently tilted westwards), or +26.3° in early October. The Sun's axis of rotation appears perpendicular to the ecliptic on January 5 and July 7. Figure 4.3 (*overleaf*) shows an example of a full-disk drawing, annotated to show B_0 and P.

In addition to yielding positional data, drawings of the solar disk can be used to assess the number of active areas. An active area (AA) is defined as a spot, or group of spots, separated from its nearest neighbour by 10° or more on the solar disk. Two spots within 10° of each other may be the products of the same local region of magnetic activity, and are accordingly counted as a single AA. Sometimes, major groups may come to span more than 10° in longitude, but these are still counted as only one AA if they contain a single major centre of activity.

At the end of each calendar month the observer can calculate the mean daily frequency (MDF) of active areas as an index of sunspot activity. This is done by summing together the AA counts for all the days on which observations were made; this total is then divided by the number of days on which observations were made. An example is given in Table 4.1 for observations made by the author in 1980 April, close to the maximum of Cycle 21.

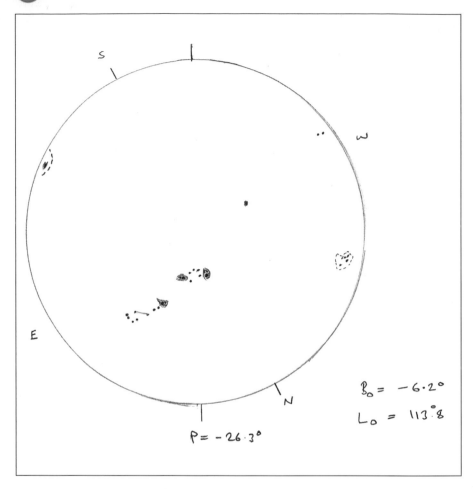

S

W

E

N

$\mathcal{B}_0 = -6.2°$

$L_0 = 113.8°$

$P = -26.3°$

Figure 4.3. Disk drawing made by projection on 1980 April 6 at 12:05 UT by the author, using a small (40-mm) refractor with a ×30 eyepiece. Two major spot groups, one near the central meridian, are particularly obvious. Faculae around spots close to the limbs are shown as dashed lines. *Drawing by the author*

As can be seen from Table 4.1, sunspot activity is quite variable from day to day; although the mean daily number of active areas is just over six, on some days there were as few as three and on others as many as eight. Also, there is considerable variation in the numbers between the two hemispheres of the Sun: early in the month the northern hemisphere was the more active, while the southern showed more activity in the latter half.

Observations in this case were quite well spread out through the month, and probably give a reasonable picture of the Sun's overall pattern of active areas in 1980 April. In several months, however, and particularly in a climate prone to spells of poor weather (such as north-west Europe), it may not prove possible for the observer to project and draw the solar disk on more

Table 4.1. Sunspot activity for 1980 April. (For comparison, figures derived from the pooled results of nineteen observers, reported in *The Astronomer* magazine, give an MDF of 6.62 (2.92 north, 3.70 south) for the same month.)

1980 April	AA Count North	AA Count South	Total AA Count
3	2	1	3
4	3	2	5
5	5	2	7
6	5	3	8
7	5	3	8
8	4	3	7
9	4	3	7
12	3	4	7
13	2	5	7
16	2	4	6
18	2	3	5
19	2	3	5
20	2	6	8
23	1	3	4
24	1	5	6
26	1	5	6
28	1	5	6
Days Observed = 17	North MDF = 45/17 = **2.64**	South MDF = 60/17 = **3.52**	Total MDF = 105/17 = **6.18**

than a couple of days. This is why it is of value to have many observers, at widely separated geographical locations, contributing their daily reports to a central collecting body such as the Solar Section of the BAA, ALPO, or the American Association of Variable Star Observers. By pooling together all the reports, the analyst will often be able to find enough data to cover every day of the month adequately, leading to a more accurate representation of the number of AAs than might be achieved by a single observer. (If, for example, observations had been possible only in the second week of 1980 April, a higher MDF would have been erroneously derived.)

Plotted against time, the MDF counts reveal quite clearly the cyclical variation in sunspot activity. Figure 4.2 presents observations from 1975 to 1998, embracing the rise of activity from minimum in the mid-1970s, through the peak of Cycle 21 in 1980, the minimum of 1986, the extended peak of Cycle 22 in 1989–90 and its minimum in 1995–96. Cycle 23 is expected to peak at the end of the 20th century, and was well on the rise by the beginning of 1998.

In addition to spots or spot groups, the observer will often notice faculae in the form of bright patches, particularly near the Sun's limb, usually around the sunspot zones but also sometimes (most often around sunspot minimum) around the polar regions. Faculae are clouds of hydrogen, slightly hotter (by about 300 K) than the photosphere and lying just above it. The commoner lower-latitude faculae often appear in regions where sunspot activity is about to break out, or has recently decayed. Some spot groups may appear immersed in faculae when near the limb. The Sun's limb darkening (caused by the increased depth of solar atmosphere through which we view the photosphere at the edges of the disk) gives the faculae increased contrast with their surroundings, making them easier to see than at the centre of the disk. Observers usually include faculae on disk drawings as dashed outlines (as in Figure 4.3) or, sometimes, shaded yellow.

Faculae may persist for some months at the site of an active area, but sunspots have a shorter lifespan. Quite often, at sunspot maximum, a well-developed group may appear on the eastern limb only to decay as it passes across the disk. Others may develop on the disk, but fail to reappear following their passage behind the western limb. As a result of the Earth's orbital motion, the Sun appears from our viewpoint to rotate on its axis once every 27 days at the equator – the synodic rotation (equivalent to a change in the longitude of the solar central meridian of 13.3° per day). So, a spot group will be visible for a maximum of about a fortnight at a time. Few survive for as many as three solar rotations. One of the largest spot groups of Cycle 22 appeared in 1991 March (Figure 4.4); flare activity associated with this group gave rise to strong aurorae (Figure 3.2).

Longitudes on the Sun are expressed in terms of the Carrington rotation period of 25.38 days originating from the work of the 19th-century English solar observer Richard Carrington, who carried out a great deal of work on the heliographic coordinates of sunspots and other features between 1853 and 1861. The first rotation in the Carrington series commenced at 12:00 UT on 1854 January 1, and solar astronomers have since used this as the zero point for longitude measurements. Annual handbooks list the Carrington heliographic longitude (L_0) of the centre of the observed disk for each day, from which longitudes of sunspots can be determined using the graduations on, for example, the Stonyhurst disks. Solar rotations are numbered in order as the Carrington series, which reached number 1935 on 1998 April 14 at 11:00 UT.

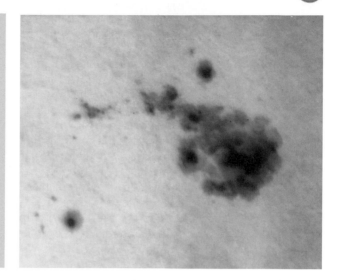

Figure 4.4. Large sunspot group on 1991 March 26, showing multiple umbrae and regions of penumbra. Flare activity in the inner solar atmosphere above this group gave rise to vigorous low-latitude auroral activity a few days earlier on March 24–25. *Photograph by Bruce Hardie, Director of the BAA Solar Section*

Sunspots are an excellent example of transient activity on the Sun which may be observed with simple equipment. Detailed daily drawings of large spot groups can show the changes in the extent and relative positions of the umbral and penumbral regions of active areas.

4.2 Monochromatic-light Observations: The Sun in Hydrogen-alpha

One of the most dramatic observations of the Sun was made by Carrington on 1859 September 1 at 11.20 a.m. While making a drawing of a large sunspot group, he noticed a brightening that appeared in the space of a few minutes, and died away quite quickly soon afterwards. This was the first recorded instance of a *solar flare*, and the event was followed a day or so later by a major geomagnetic storm and extensive auroral displays. Only very rarely do solar flares become visible in white light, and an observer would be extremely lucky to make such a sighting in a lifetime of solar work!

Such activity is much more likely to be seen if the Sun is monitored in the isolated wavelength of hydrogen-alpha (H-alpha) light at 656.3 nm. During a total solar eclipse (Section 4.3) the innermost layer of the Sun's atmosphere – the *chromosphere* – is seen as a narrow ring of red light, shining at the wavelength of H-alpha, surrounding the dark body of the Moon while the brilliant photosphere is hidden. The chromosphere forms a layer about 10 000 km deep overlying the photosphere, and has a temperature of about 10 000 K.

At some total eclipses, higher extensions of chromospheric material – *prominences* – may be seen reaching into the corona. These prominences are loops of gas, held in place by magnetic field lines emerging from the solar interior. Gas in prominences is maintained at lower temperatures than the extremely hot (1 million K) surrounding corona by the insulating effect of the magnetic fields that shape them. Prominences can reach as high as 50 000 km above the solar surface.

Prominences can be observed at times other than during total eclipses by making use of a narrow-passband filter, which must be securely mounted over a telescope's objective. Such a filter, which should be bought only from a reputable supplier, will exclude virtually all of the Sun's light except that close to the H-alpha wavelength. This will certainly dim the Sun, but many observers also like to take the precaution of adding a safety filter of Mylar.

H-alpha filters open up a whole new area of solar activity to the observer, but are by no means cheap. The narrower the passband, the higher the price, but the better the contrast in the telescopic view. Filters may need adjustment, usually by a tilting mechanism, to achieve the best performance during observation. Heating by the Sun during an observing run may detune a filter. Some of the more expensive systems use a self-contained thermal unit to keep the filter at its optimal working temperature. H-alpha filters will gradually deteriorate with use. For the dedicated solar enthusiast, however, the price is worth paying.

Viewed through an H-alpha filter, the Sun will show prominences around the limb on most days. These may change in appearance, slowly or rapidly. Some *quiescent prominences* are long-lived, and may persist for several months with only minor changes from day to day. Obviously, quiescent prominences will be seen from a different angle from one day to the next as the solar rotation carries them over the eastern limb onto the disk, or over the western limb and out of view.

Eruptive prominences show more rapid changes, on timescales of tens of minutes. Some may eject their material out into the corona and disappear. Such events are most likely to be seen around sunspot maximum, when the chromosphere is often disturbed by violent activity associated with solar flares.

Hydrogen-alpha filters also allow the chromosphere over the visible hemisphere of the Sun to be observed as a disk. Seen against the disk, prominences appear by contrast as dark lines termed *filaments*. These can be seen only under the best of conditions, however, with a carefully tuned narrow-passband filter: the presence of atmospheric haze or cirrus cloud will render filaments invisible, and also compromise viewing of prominences at the limb.

Filaments, being identical to prominences (simply viewed from above), are subject to changes in appearance caused by the same mechanisms. Of particular interest are *disappearing filaments*, whose material is ejected into the corona. Sometimes material from a disappearing filament will arrive in near-Earth space, giving rise to increased geomagnetic/auroral activity; the major, global auroral storm of 1991 November 8/9 had such an origin.

The underlying causes of prominence eruptions or filament disappearances are magnetic. A principal source of chromospheric disturbances is flare activity in the inner solar atmosphere above sunspot regions. Here, intense magnetic fields of opposite polarity may be brought into close proximity, resulting eventually in *reconnection* and the explosive release of vast amounts of energy. As discussed earlier, flares may actually result from the release of magnetic stress still higher in the solar atmosphere, associated with coronal mass ejections (Section 3.1.1).

Shock waves expanding outwards from solar flares will certainly disturb filaments, giving rise to rapid motions of gas within them as filamentous *surges*. In severe cases, a filament (or prominence) may become detached from its base in the chromosphere, leading to its ejection into space. Some hours later, the solar magnetic field in the region may return to roughly its previous configuration, and a prominence or filament will again form in more or less the same location.

Suitably equipped observers make drawings of prominences as seen at the limb each day. Series of drawings made at intervals of a few minutes during the eruption of a prominence can provide an interesting "time-lapse" portrayal of more vigorous activity (Figure 4.5, *overleaf*). Daily counts of prominences are made by some observers, allowing a prominence MDF to be derived in exactly the same way as for

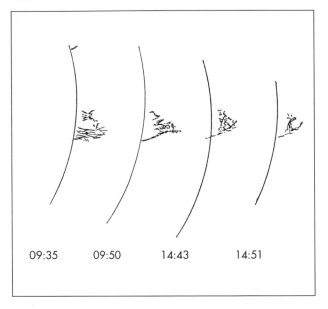

09:35   09:50   14:43   14:51

Figure 4.5. A series of drawings showing the development of a semi-active solar prominence on 1997 May 22. (Times are UT.) *Drawing by Bruce Hardie*

sunspot active areas (Section 4.1). The prominence MDF usually reaches its maximum during the rising phase of the sunspot cycle.

Observations at H-alpha wavelengths make it possible to monitor for flare activity itself. Certainly, the chances of seeing a flare at these wavelengths are much greater than in white light. Flares in H-alpha appear as localised brightenings in the chromosphere above active regions: the projected white-light image will prove a useful guide as to where to watch. In H-alpha the underlying spot group may be hard to recognise, perhaps appearing only as small, localised darkenings.

Flares occur most often in the years around sunspot maximum (hardly surprising, given that they are triggered by interactions between regions of high magnetic flux emerging in the sunspot zones). At maximum, as many as twenty-five flares have been recorded each day by professional astronomers carrying out constant monitoring programmes.

Typically, a flare may last from 20 minutes to a couple of hours. There is no need to watch the H-alpha Sun continuously to see such events. A better approach, recommended by experienced observers, is to check every 20–30 minutes throughout the day. If a brightening is seen in progress, subsequent developments can be followed. As with other transient phenomena, the observer will require patience: several clear days may pass without a single flare being seen. Chances are certainly highest

Table 4.2. Solar flares

Class	Area (Millionths of a Solar Hemisphere) Affected
Subflare (S)	<100
1	100–250
2	250–600
3	600–1200
4	>1200

Subclasses: f faint; n normal; b bright. Thus Sf is a very weak event, 2n fairly strong, 4b extremely bright.

when sunspots are numerous, and the best candidates for flare sites are large spot groups with multiple umbral centres, indicative of complex emerging magnetic fields. Flares occurring above such groups are often followed, as in 1989 March and 1991 March (Section 3.1), after a day or so, by low-latitude auroral activity. Sightings of flares close to the central meridian can sometimes provide a useful alert to the possibility of subsequent enhanced auroral activity.

Flares vary considerably in intensity. One means of estimating the intensity is to express the area they affect in terms of millionths of the visible solar hemisphere. (Similar size estimates can be made, with the use of suitable graticules, for sunspot groups.) Table 4.2 summarises the commonly used scale. In general, the bigger the flare, the greater the chance of a geomagnetic disturbance if all the other circumstances – alignment with Earth and later coupling between interplanetary and terrestrial magnetic fields – are favourable.

4.3 Solar Eclipses

Among the most eagerly followed transient astronomical phenomena are eclipses of the Sun, when the Moon passes in front of the Sun and blocks off part or all of its light. Strictly speaking, such events are actually *occultations* (Section 5.2). Since solar eclipses occur with the Moon in front of the Sun, it follows, of course, that they occur at new Moon.

The eclipses that attract the most interest are total eclipses of the Sun, which come about through the happy coincidence that the Moon has a diameter about 1/400 that of the Sun, which lies about 400 times farther away. Total solar eclipses are rare events, however, at any one geographical location, and in the last hundred years a thriving market in "eclipse tourism" has grown up. The chance to take a holiday in a distant country, coupled with the brief splendour of a total solar eclipse, is a great attraction for those who can afford it. Not all total events are ideally placed: on 1997 March 9, for example, several observers made the arduous journey to the remote frozen wastes of Mongolia only to be greeted with clouds, while the 1991 July 11 eclipse from the ostensibly very favourable location of Hawaii was also clouded out for many. Tales of frustration abound, and serve as a reminder that observation of transient astronomical phenomena, no matter how predictable, can often depend on luck as much as anything.

For those not in a position to travel, *partial solar eclipses* are more common, if less spectacular, at a given location. The line of totality during total eclipses is generally narrow, but observers located to either side of it will see a partial eclipse, of diminishing extent with increasing distance from the line. Not all eclipses are necessarily total; in some, the line of totality may miss the Earth.

During a partial eclipse (Figure 4.6) the dark body of the Moon will initially make a small "nick" on the Sun's western limb (the moment of first contact), gradually intruding further across the disk as the Moon travels eastwards. Some partial eclipses are mere grazes: the author observed an eclipse in 1983

Figure 4.6. Partial solar eclipse, 1971 February 21. *Photograph by the author*

December from Edinburgh, for example, in which only 3% at most of the Sun's southern limb was obscured, for just a few minutes.

The maximum extent of a partial solar eclipse is expressed as the percentage of the Sun's disk which is obscured. Even quite a large partial eclipse – up to about 80% – will have little effect on the quality of sunlight. Without looking at the Sun's eclipsed disk through a suitable, safe filter, a casual observer might be unaware that such an event was in progress. Sometimes more noticeable is a drop in temperature. The partial eclipse of 1996 October 12 was fairly large at maximum for observers in the British Isles, reaching in excess of 60%; many commented on the marked temperature decrease during the event, which happened on an otherwise balmy autumn Saturday afternoon.

The best way to view a partial solar eclipse is to project the image onto a piece of white card, exactly as for sunspots (Figure 4.1). Under good conditions the irregular profile of the Moon's limb may be apparent in the projected image. Also apparent is just how dark the body of the Moon is in comparison with any sunspots present on the disk, again emphasising that spots are dark only by contrast with the surrounding photosphere. Eclipse "viewing glasses", made from Mylar filters, are also useful for observing partial eclipses, and can be obtained from most astronomical equipment suppliers.

The Moon moves eastwards by about its own diameter in an hour, so partial eclipses can last up to a couple of hours from first contact to when the Moon moves off the Sun's eastern limb at last contact. The precise duration depends principally on how extensive the eclipse is at maximum: near-grazing events will be much shorter than those where the Moon takes a substantial "bite" out of the Sun.

Total solar eclipses are regarded by many as the most spectacular of all astronomical phenomena. Following first contact, the partial phase grows ever larger until second contact sees the Moon completely cover the photosphere, extinguishing the Sun for those within the shadow cast on the Earth. In the last few minutes before totality there is a marked change in the quality of sunlight, particularly once more than 95% of the Sun is obscured. Shadows cast by the thin remaining crescent become very sharp, and the air temperature drops. As the Moon nears second contact, the last chinks of sunlight shining through valleys on its irregular limb give

Figure 4.7. One of the most dramatic phenomena during a total solar eclipse is the diamond ring which immediately precedes and follows totality, seen here from Curaçao during the Caribbean eclipse of 1998 February 26, and captured with a 400-mm telephoto lens. *Photograph by Pam Spence*

rise to a partial ring of intense bright spots, named Baily's beads after the 19th-century English astronomer Francis Baily, the first to describe them in detail. Sometimes there is just a single intense spot of sunlight which, combined with a faint glow around the disk of the Moon, gives the beautiful "diamond ring" effect (Figure 4.7). After just a few seconds, Baily's beads vanish and the darkness of the Moon's shadow finally sweeps across as totality begins.

During totality, it becomes possible to see the Sun's inner atmosphere as the bright, pearly *corona* surrounding the dark body of the Moon (Figure 4.8). The corona is comparable in brightness to the light of the full Moon (so even at mid-totality the sky is not completely dark) and may extend for several solar radii out from the Sun. Depending on the stage in the sunspot cycle, the corona may be dominated by long equatorial steamers with little material at the poles (sunspot minimum), or may have a fairly symmetrical, even distribution around the disk (sunspot maximum). Binocular examination will reveal fine structure in the corona, resulting from the intense magnetic fields in the inner solar atmosphere, including loops or "helmets", and longer streamers. Care should be taken,

Figure 4.8. Total solar eclipse, 1998 February 26, showing the Sun's inner atmosphere, the corona, as seen from Curaçao in the Caribbean. *Photograph by Pam Spence*

obviously, to avoid binocular viewing close to the return of Baily's beads at third contact as the Moon's trailing limb leaves the disk.

As well as revealing the corona, which is never the same from one eclipse to the next, totality allows us to view the chromosphere, which appears as a red ring shining in the light of H-alpha (Section 4.2). The most obvious chromospheric features will be prominences, extending some way beyond the lunar limb. Since the Moon does not normally have exactly the same angular diameter as the Sun at eclipses, the chromosphere is often seen better at one limb than the other. Just after second contact and the onset of totality, the chromosphere above the Moon's leading limb is seen to best advantage. Towards totality's end, the trailing limb reveals the chromosphere on the Sun's western side, while that at the eastern side is hidden.

All too quickly, totality's end is heralded by sunlight shining through depressions on the Moon's limb, as either the diamond ring or Baily's beads. The partial stages are then repeated in reverse until, about an hour later, the Moon slips off the Sun's disk.

Totality can have a maximum duration of $7^m\ 31^s$; at most eclipses it is rather shorter, but by choosing an appropriate eclipse the willing and able traveller has the chance to take in totality lasting several minutes. There are several schools of thought as to what the observer should do during these few precious minutes of totality. The simplest approach is to sit back and enjoy the spectacle, which will remain in the memory for ever. Others aim to record the precise moments of the beginning and end of totality, with a view to measuring the Sun's angular diameter and/or the Moon's motion. For many, the principal target is photography of the eclipsed Sun's corona.

The choice of lens and film for eclipse photography depends very much on what the observer wants to record. The veteran British eclipse-chaser Mike Maunder advocates using a wide-angle (28-mm, $f/2.8$) lens with ISO 400 film to record the general scene at the eclipse site during totality, with exposures of a few seconds. Particularly if the eclipsed Sun is not too high in the sky, the inclusion of the horizon and foreground features adds to the composition.

Exposures of 1 second or so with a standard 50-mm $f/2$ lens will give a larger image of the Sun and surrounding sky, which may include brighter zodiacal stars (Regulus or Spica, for example) in the background, and planets (particularly Mercury and Venus). Sooner or later, another "eclipse comet" to compare to that of 1948 – previously unknown, and revealed only when close to the Sun during the relatively dark window of totality – will surely be found and photographed.

Closer in to the eclipsed Sun, observers wishing to record the corona or prominences will need to use a telephoto lens. Mike Maunder recommends slower (finer-grained) ISO 100 film for recording prominences and faster (ISO 400) film for the corona. The best advice is to be ready to take a range of exposures during totality, bracketing by several stops (Covington, 1991). On ISO 400 film, with a 135-mm focal telephoto lens, the outer corona should be recorded in exposures at $f/8$ bracketed around 1/8 second. The red-sensitivity of some colour emulsions particularly favours the recording of prominences.

In addition to the brief visibility of the corona and prominences, there are other phenomena to watch out for at a total eclipse. The onrushing shadow of the Moon (travelling at about 2500 km h^{-1}) may be apparent on the distant landscape a minute or two before totality, and will be seen afterwards heading rapidly away to locations farther along the track.

The enigmatic *shadow bands* may also be seen on the ground, or on light-coloured surfaces, just before and just after totality. A particularly marked display of these closely spaced ripples of light and dark, a matter of a few centimetres broad, was seen at the Caribbean total eclipse of 1998 February 26, lasting for about 30 seconds on either side of totality. Shadow bands are believed to result from differential refraction of the thin sliver of sunlight at the Moon's limb close to totality by air masses at slightly different temperatures. They seem to be more prominent at some eclipses than others.

Not only is the Moon's orbit inclined relative to the ecliptic, it is also somewhat elliptical. The Moon appears larger (angular diameter 33′ 31″) at perigee, its closest to Earth, than at apogee (29′ 22″). These differences are not readily apparent to the naked eye, but are significant in terms of eclipses. Eclipses where the Moon passes centrally over the Sun while close to apogee cannot be total. Instead, a ring – or annulus – of sunlight remains throughout, and the corona does not become visible. Such *annular eclipses* are impressive in their own way, but are not as spectacular as total eclipses. The annular phase, with the Moon entirely superimposed on the Sun's disk, can last for up to $12^m 30^s$.

Table 4.3 lists forthcoming solar eclipses until 2003. Among the most eagerly awaited of these is the total eclipse of 1999 August 11, which will be the first visible from mainland Europe for some time. The path of totality will also cross the south-west tip of England. An annular eclipse on 2003 May 31 will be visible from northern Europe, including the Shetland Isles, and from Canada.

Table 4.3. Solar eclipses 1999–2003

Date	Area of Visibility	Type
1999 February 16	Indian Ocean	Annular
1999 August 11	SW England, Europe, Middle East	Total
2000 February 5	Southern Ocean	Partial
2000 July 1	Southern Ocean	Partial
2000 July 31	Russia, N Europe, Canada	Partial
2000 December 25	US, N Atlantic	Partial
2001 June 21	Africa	Total
2001 December 14	E Pacific	Annular
2002 June 10	Pacific, S America	Annular
2002 December 4	Africa, Indian Ocean	Total
2003 May 31	Middle East, N Europe, Shetlands, Canada	Annular
2003 November 23	Southern Ocean	Total

References and Resources

Covington, M, *Astrophotography for the Amateur*. Cambridge University Press (1991).

Maunder, M and Moore, P, *The Sun in Eclipse*. Springer (1998).

NASA Eclipse Web Site: http://planets.gsfc.nasa.gov/eclipse/eclipse.html

Nicolson, I, *The Sun*. Mitchell Beazley (1982).

NOAA Current Activity Web Site: http://www.sunspot.noaa.edu/index.html

Philips, KJH, *Guide to the Sun*. Cambridge University Press (1992).

Wentzel, DT, *The Restless Sun*. Smithsonian University Press (1989).

Chapter 5

The Moon

As the Earth's nearest neighbour in space, at an average distance of 384 400 km, the Moon has long been viewed as the ideal object for beginners. Even binoculars or a small telescope will allow lunar features to be resolved in reasonable detail, and many good maps are available for guidance (e.g. Rükl, 1991; Hatfield, 1968). The Moon presents an essentially unchanging face, and its captured rotation means that only 59% of the lunar terrain is visible from our viewpoint on the Earth. The present-day moonscape – the relatively smooth, dark lava plains (maria) and the lighter, crater-scarred highland regions – with only a few comparatively recent crater additions (such as Tycho and Copernicus, estimated to be 100 million and 810 million years old respectively), dates back more or less to the end of the late stage of bombardment by small planetesimals around 3800 million years ago.

Many amateur astronomers start their observing careers by becoming familiar with the apparently unchanging surface of the Moon. A large number of regular lunar observers continue to draw and photograph the Moon's features, even in this post-Apollo age. Although the Moon has been orbited, walked on and directly sampled geologically, the mapping of some regions – especially near the poles – is incomplete, and dedicated observers find value in trying to see these areas more clearly and record them when conditions are favourable.

The Moon is one of the few observational subjects which is unaffected by light pollution (indeed, as a *source* of unwanted light in the sky, it is cursed by meteor observers, deep-sky enthusiasts and others who find their activities curtailed around full Moon). Observations are easily carried out from urban locations. Opinions vary on what type of telescope is ideal for lunar (and planetary) study. In general, lunar work can be done effectively with reflectors of 150 mm aperture upwards, or refractors of 100 mm aperture or greater. A 150-mm reflector can theoretically resolve features as small as 2 km in diameter on the Moon's surface.

Craters and mountains are generally best seen when close to the day/night line of the *terminator*, where shadows throw features into strong relief. Around full Moon, when shadows are virtually absent, the ray systems emanating from young, bright craters such as Tycho are seen to best advantage, as are subtle shading differences in the dark maria.

The principal limitation on the resolution of lunar detail for most observers is imposed by atmospheric turbulence. It is the unsteadiness of the air that causes the marked twinkling of stars on frosty nights, as pockets of air at different temperatures (and, therefore, of different refractive indices) pass across the observer's line of sight. Observers describe the stability of the atmosphere as *seeing*. Seeing should not be confused with *transparency*, the clarity of the air. Indeed, some of the best, steadiest seeing conditions are found on calm, hazy nights when the stars are dimmed. Both lunar and planetary observations are affected by seeing conditions, which should always be indicated on observational reports, using the standard Antoniadi scale:

I   Perfect seeing, without a quiver.
II   Slight undulations; moments of calm lasting several seconds.
III Moderate seeing, with larger air tremors.
IV   Poor seeing; constant troublesome undulations.
V   Very bad seeing; even a rough sketch impossible.

5.1 Transient Lunar Phenomena

In the past, some lunar features have been suspected of changing their appearance over periods of decades. Perhaps most notably, the 10-km diameter feature Linné on the western side of Mare Serenitatis was depicted on a map by Wilhelm Beer and Johann Mädler in 1843 as a deep pit, but has been seen since 1866 as a white, hazy ring with only a small central craterlet. The occurrence of any real changes is, however, doubted by most serious lunar scientists.

More time and study has been devoted to *transient lunar phenomena* (TLPs, in North American parlance LTPs), which are occasionally reported by those who observe the Moon regularly. TLPs may take the form of coloured (usually reddish) glows, obscurations or changes in the apparent brightness of features. Such changes are, naturally, most likely to be reported by observers familiar with the normal aspect of regions of the Moon's surface. As the name suggests, TLPs are of short duration. Typical reported lifetimes range from 10–20 minutes to a couple of hours. Observations of suspected TLPs have been collected by the BAA Lunar Section since the 1960s, and by ALPO since the 1970s. In all, about 1500 separate instances have been logged.

Perhaps the earliest recorded putative TLP was an event described by monks at Canterbury on the evening of 1178 July 18, consisting of an apparent glow near the upper cusp of the waxing crescent Moon, followed by an overall darkening. It has been suggested that this event may have been associated with the impact that formed the young crater named Giordano Bruno. In common with many more modern reports, this observation is regarded with a degree of healthy scepticism by researchers. Among other candidates for historical TLP observations are reports of red glows by William Herschel in 1783 and 1787.

Over a great many years, further reports of such glows have been made, mainly by amateur observers. But the study of TLPs was given considerable impetus by observations made using the 1.2-metre reflector at the Crimean Astrophysical Observatory by professional astronomer Nikolai Kozyrev on 1958 November 3. He observed a reddish glow over the central peak of the 120-km diameter crater Alphonsus, which with Ptolemaeus and Arzachel forms part of a distinctive crater chain near the centre of the Moon's visible disk. The glow persisted for about 30 minutes, during which time Kozyrev was able to obtain a spectrum – the only instance when this has proved possible for a TLP – which appeared to show evidence for emission lines of carbon. Doubts have since been expressed over the reliability of the spectrum, and the observation remains controversial (Doel, 1996).

Whatever the reality of Kozyrev's observation, analysis of the records suggests Alphonsus to be one of the most favoured sites for alleged TLP activity. Others include the dark crater Plato just north of the huge Mare Imbrium basin, and the craters Copernicus, Proclus, Theophilus, Eratosthenes, Gassendi and Bullialdus. Some TLP reports cluster around the rims of Mare Imbrium, Mare Crisium and Mare Serenitatis. The prime location, accounting for about a third of all reports, is Aristarchus.

The 40-km diameter, 3-km deep crater Aristarchus is the brightest feature on the visible lunar hemisphere (Figure 5.1), appearing brighter than the surrounding terrain, even when seen by earthshine (Section 5.2) on the unilluminated part of the Moon before first quarter. While the vast majority of the craters in the lunar highlands are indubitably of impact origin, a few – Aristarchus among them – show convincing evidence of association with regions where volcanic activity has

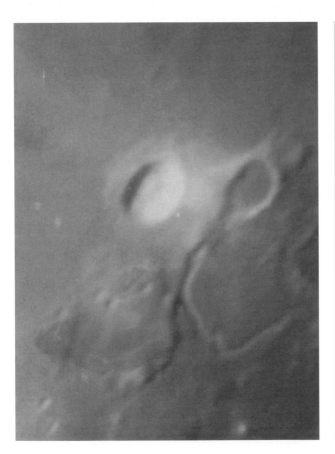

Figure 5.1. The Aristarchus region of the Moon, regarded by some observers as a TLP "hotspot", photographed on 1982 December 27 at 21:30 UT using a 355-mm Cassegrain telescope at f/80 for a 0.25-second exposure on Ilford XP1 ISO 400 film. *Photograph by Martin Mobberley*

taken place. The plateau on which Aristarchus sits, above the plain of Oceanus Procellarum, is crossed by sinuous rilles that look as if they were produced by localised lava flows. Other features indicative of past volcanic activity on the Moon are chains of small craters which appear to be associated with fault lines, and clusters of low domes.

Observation of TLPs requires familiarity with the features of the Moon. This demands a long, personal programme of repeated observation, and many hours at the eyepiece. Lunar observers of long experience advocate observing and drawing craters and other features under various solar illumination conditions (Moore, 1974). Time should be spent recording on paper a feature's appearance under morning (waxing), full and evening (waning) solar illumination. It is best to concentrate on one relatively small region at a time.

Having learned their way around the Moon, seasoned observers should immediately be able to see anything unusual. Checks should be made on every clear night, paying particular attention to locales for which TLPs have been reported, but not neglecting the rest of the surface. On the vast majority of nights, of course, nothing out of the ordinary will be seen. Should a TLP be suspected, a record of the time and appearance – preferably at least a sketch – should be made.

Some observing bodies operate TLP alert networks. If unusual activity is suspected, a telephone alert can be passed to other observers. Care should be taken not to provide too many specific details of the putative TLP, because this might then bias what others will report. Ideally, events require independent confirmation by observers who are broadly unaware of what may have been seen by others.

The commonest reported form of TLP is a reddish glow affecting a small area, perhaps up to 15 km in diameter. These glows can be emphasised by the use of filters. Several lunar observers employ an alternating filter system, called a Moonblink, at the eyepiece to check possible TLPs. The Moonblink uses Wratten gelatin-based filters, which are also useful in planetary work. A blue number 44A filter is usually coupled with a red number 25 filter in this system. Viewed through the blue filter, a reddish glow will appear to have increased contrast with its surroundings, while the red filter will make it appear dim.

The other forms of TLP are harder to define. Changes in the apparent brightness of many features

may as easily be ascribed to unfamiliar solar illumination angles as to any real, intrinsic effect. Obscurations
and coloration may stem from atmospheric seeing conditions rather than activity on the Moon's surface.
Coloured fringes around features resulting from
optical problems can be eliminated rapidly from sus-
picion by checking for their presence around other
features. Putative obscuration events in Plato have
frequently been dismissed as resulting from variable
seeing conditions affecting the visibility of small crater-
lets on the floor of the main crater.

For any possible TLP activity, it is critically important to make an objective report. The report should give
details of the phenomenon or phenomena observed.
Meteorological conditions at the time of observation
should be noted – the presence of thin, high cloud or
haze, for example, may be a root cause of some col-
orations. The seeing should be recorded, using the
Antoniadi scale (page 70), with the observation.

Naturally, many are sceptical of the reality of transient lunar phenomena. A major problem is finding a
mechanism by which TLPs can occur on the Moon
in the current epoch, given its level of geological
(in)activity. Intense study, culminating in the Apollo
landings of the late 1960s and early 1970s, has built up
a picture of an essentially dead body. The lunar crust
has a depth of 600–800 km, much thicker than that of
the Earth (French, 1977). Certainly, the top 60 km or so
of the crust appears to consist of solid rock, with a thin
outer layer of 1 km depth consisting of material shattered by meteoritic impacts. There is no evidence for
magma chambers or other similar reservoirs of volatile
material as found on the Earth in regions of volcanic
activity, or for tectonic processes associated with
crustal movements. The Moon's volcanoes became
extinct a very long time ago.

While the margins of the lunar maria basins have
been proposed as lines of weakness from which gas
may emerge from the Moon's interior, these features
are ancient, dating back to the upwelling of lava 3000
million years ago. There has been ample geological
time for the relief of any stress in these regions. Deep-
seated moonquakes have been detected, originating
from the base of the crust at a depth of 600–800 km, by
the seismometer packages left behind by the Apollo
astronauts. There is some correlation between these
events and the orbital position of the Moon: they
appear to be commonest around perigee, when the

Moon is closest to the Earth. It is difficult, however, to reconcile such activity with TLPs at the lunar surface.

Popular theories of TLPs suggest that localised obscurations may have their origin in the escape of trace quantities of gas trapped in the lunar interior, picking up fine dust on the way which then spreads out as a small cloud over the Moon's surface. Fluorescence of the gas and/or dust in sunlight may account for glows, or temporary regional brightenings.

The reality, or otherwise, of TLPs remains a matter of some debate. Rather like those who maintain the reality of sounds made by meteors, most observers who have reported them are firmly convinced of the reality of TLPs. Ultimately, more observations and rigorous reporting are required to resolve the issue, and several lunar observing organisations are happy to continue the collection of objective reports as part of their programmes.

5.2 Lunar Occultations

During its monthly orbit around the Earth (taking 27.6 days – the sidereal month – to return to the same position with respect to the stars), the Moon repeatedly passes in front of stars, temporarily obscuring them from view. These *occultation* events are of interest both to a hard core of amateur astronomers who observe them regularly, and to professional astronomers studying the dynamics of the Earth–Moon system.

Lunar occultations are transient phenomena *par excellence*. At the telescope eyepiece, the observer watches intently as the Moon's limb advances on a star; then, in an instant, the star is gone. The absence of an atmosphere around the Moon means that disappearances are immediate and sudden – literally, a case of blink and miss it.

The aim in occultation observing, from a scientific viewpoint, is to make accurate timings. Most observers prefer to cover *disappearance* events on the Moon's leading edge, which are easier to time, since the star is visible up until the moment of contact. However, there is also much value in observing *reappearances* on the Moon's trailing limb, and such events are targets for the skilled and dedicated.

The Moon's orbit is inclined at an angle of 5° 8′ 42″ to the ecliptic plane, which it crosses twice at the nodes. The line of nodes, connecting the ascending and descending nodes (which, obviously, lie 180° apart) gradually moves westwards over an 18.6-year period known as the Metonic cycle. From one year to the next, each node appears 19° 21′ further to the west along the ecliptic. A consequence of this so-called regression of the line of nodes is that the Moon's apparent path against the star background gradually changes from year to year. Over the course of a Metonic cycle there will be only certain favourable periods during which it is possible to observe occultations of particular stars. For example, series of occultations of Aldebaran and the Hyades stars occurred from 1977 to 1981, then again from 1996 to 2000; through the rest of the cycle, the Moon missed these stars. There are similar "occultation seasons" for the Pleiades and for Antares.

In theory, all stars within 6° 37′ of the ecliptic can be occulted. Stars with high ecliptic latitudes, such as Aldebaran and Antares (+5° 47′ and 4° 56′ respectively) have only a single interval during which they can be occulted; those closer to the ecliptic, such as Regulus (ecliptic latitude +0° 46′), have two intervals during which they may be occulted, separated by about nine years.

5.2.1 Observing Occultations

A principal requirement for the successful observation of occultations is a list of predictions indicating when they are expected to occur. The time of the event will depend on the observer's exact geographical location. Such predictions are approximate to some extent: part of the scientific value of the observations is in determining the actual time of the event, refining the Moon's *real* orbital motion relative to its expected, modelled motion.

As mentioned earlier, stars within 6° 37′ of the ecliptic plane can be occulted, and for many years the standard listing of such stars was the *Zodiacal Catalogue* (ZC), published by the US Naval Observatory in 1940. The ZC listed 3539 stars brighter than mag. +8.5. More recently, a combination of star positions from the ZC, SAO and AGK3 catalogues has been used to generate occultation predictions. The high-precision positional data generated by the Hipparcos satellite between 1989 and 1993 have become the standard source for star positions used in calculating predicted times for lunar occultations.

For general purposes, lists of occultation predictions for standard locations are given in annual publications such as the BAA *Handbook* and the Royal Astronomical Society of Canada's *Observer's Handbook*. North American predictions are given for 17 locations. The BAA publishes predictions for Edinburgh and Greenwich in the UK, Melbourne and Sydney in Australia, and locations in New Zealand and South Africa. An overview of the year's more interesting events for observers in North America is usually presented in the magazine *Sky & Telescope* (see e.g. Dunham, 1998).

The BAA *Handbook* and other more general reference sources list occultation predictions for stars down to about mag. +7.0, sufficient for the needs of occasional observers. The occultations that attract the most attention from more casual observers are those of bright stars – notably Aldebaran, Regulus, Spica and Antares – or the planets Venus, Mars, Jupiter and Saturn. More serious observers, who set out to undertake extensive sets of timings, can obtain lists of occultation predictions from local representatives of the International Occultation Timing Association (IOTA).

The equipment required for occultation observing is reasonably simple. At a pinch, bright events can be covered using a pair of 10×50 binoculars, firmly mounted on a sturdy tripod. Better results can be obtained with a small telescope; portable, altazimuth-mounted instruments used with a magnification of ×50 are perfectly adequate. A stopwatch readable to a precision of 0.2 second or better is essential for obtaining accurate timings. Analogue watches remain useful, but many observers have taken to using digital stopwatches, with which there is less chance of ambiguity in the final readout. Access to a reliable central time source is also important. Many observers use national "speaking clock" telephone time services; broadcast radio time signals are also useful.

The most favourable events for occultation observers to cut their teeth on are disappearances on the dark limb of the waxing Moon before full. Particularly favourable are those events in the first week or so of the lunation following new Moon, when the bright crescent may be accompanied by *earthshine* – the reflection of sunlight from the Earth onto the hemisphere of the Moon unilluminated by the Sun, traditionally referred to as the "old Moon in the new Moon's arms". Earthshine is the observer's ally, revealing in its pale, greyish light the limb of the Moon as it advances on the target star. After first quarter, earth-shine is very much reduced and it is harder to make out just where the Moon's dark limb is, relative to the star. Occultations during the Moon's crescent phase also have the advantage of being observable during the evening hours.

Successful observation demands some advance planning, starting with a check in an appropriate listing for events likely to be visible, and their approximate timing. As mentioned above, the predictions given in such sources as the BAA *Handbook* are for "standard" centres, and unless the observing site is very close to one of these, some corrections will have to be made to allow for the geographical difference, using tables provided in the source. It is worth doing the necessary arithmetic a few days ahead of the event, and being prepared to have the telescope set up at least 20 minutes before the expected occultation.

As an aid to locating the star relative to the contact point on the lunar limb, predictions list the *position angle* at which the occultation should occur. Position angle (PA) is given in degrees, measured anticlockwise from the lunar north pole (000°), through 090° on the leading limb (easternmost on the sky) at right angles to the pole, 180° due south, and so on. Some sets of predictions also give the *cusp angle* at which the event is expected to occur. Since the Moon is inclined with respect to the ecliptic, its poles do not necessarily correspond to the tips of the cusps (the "horns" of the crescent Moon), as shown in Figure 5.2.

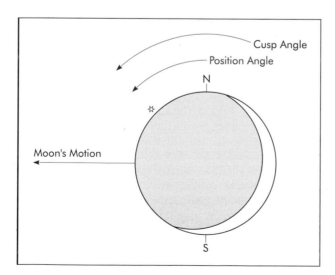

Figure 5.2. Position angle (PA) and cusp angle (CA) for a lunar occultation. In the particular instance illustrated here, the event will be a disappearance behind the dark, easterly limb of the waxing crescent at PA 042°, CA 067°. Reappearance will occur on the bright limb at PA 318°.

A reasonably long lead-time allows the observer to become "field-adapted" at the eyepiece, growing used to the gradual advance of the Moon towards the star. The Moon moves eastwards at a rate of roughly 30′ (about its own diameter) per hour. Watch very carefully, stopwatch in hand, over the last couple of minutes leading up to the expected time of the occultation, and be ready to start the stopwatch the instant the star disappears behind the lunar limb. Concentration at this point is vital!

The next stage of the observation, once the star has disappeared and the watch is running, is to obtain the timing. This is achieved by accessing the standard time signal and, at a suitable moment, stopping the watch. The time of the occultation event can then be found by simple arithmetic. For example, following an occultation the watch might be stopped precisely on the broadcast time signal for $21^h 10^m 00^s$ UT, having been running for $4^m 23.2^s$. By subtraction, this gives a timing of $21^h 05^m 36.8^s$ UT.

There are, naturally, limits to the accuracy that can be achieved by using a stopwatch. Among the limiting factors are the time required for the watch's mechanism to respond to the observer pressing "start", and running errors (the watch may run fast or slow, over long time intervals). The latter can be minimised by taking as little time as possible between observing the occultation and obtaining the time signal. Modern digital watches are less prone to running errors than their mechanically driven analogue counterparts, but over short timescales there should be little practical difference in performance between the two. Watch errors can conveniently be assessed relative to radio or speaking clock time marks on cloudy nights, or by daytime, and noted for later reference.

Less easy to assess are errors due to the observer's physiological reaction time – the *personal equation*. Some attempts have been made over the years to test observers' reactions using simulated occultations on personal computers, where software controls the (randomised) disappearance, and records the time taken for the observer to respond. Experienced observers are reckoned to have reaction times of 0.1 second under these conditions; however, there are substantial differences between the comfortable indoor environment where such tests are usually conducted, and the cold or windy conditions often encountered during actual observing runs. A more reasonable estimate of a typical observer's personal equation may be 0.2–0.5 second.

One interesting point to emerge from research on reaction times, presented by Geoff Kirby at a BAA meeting in the late 1980s, is the effect of alcohol. Even a single glass of wine had a markedly deleterious effect! Indeed, the most dramatic deterioration in reaction time occurred, in the test subjects, over the first couple of glasses. The moral must surely be that alcohol is to be avoided before conducting occultation observations.

Timing of disappearance events is reasonably straightforward. Much more demanding – and prone to error – is the timing of reappearances from the Moon's trailing limb. There is little point in attempting to time reappearances for faint stars on the Moon's sunlit, bright limb. Events taking place at the dark limb of the waning Moon, especially more than 3 or 4 days after full, are more likely to be effectively observed. For these events it is essential to have a good idea of the position angle on the lunar limb where the star is to reappear. The observer's attention is then concentrated on this point, and the stopwatch is started the moment the star springs into view. For bright stars such as Aldebaran, reappearances can be spectacular.

To be of scientific value, an occultation observation needs to be reported to one of the central collecting agencies. IOTA collects reports from observers around the world, and is the main clearing-house for North American observers. Some of the national organisations, such as the BAA and ALPO, have dedicated occultation groups within their lunar observing sections which collate members' reports for forwarding.

Reports should be submitted reasonably soon after observing an event. Most observing bodies issue standard forms on which to log details. Of equal importance to the timings themselves are other items of information, including the instrument(s) used and a precise geographical location for the observation site (accurate to within 30 metres in latitude and longitude, and 10 metres in altitude above mean sea level). The report should also give an indication of accuracy, based on the observer's personal equation and previously determined stopwatch errors. Some timings will be more reliable than others, depending on weather and other conditions, which should also be recorded. An example from the author's observing log is given in Table 5.1.

Table 5.1. Sample occultation observations

Observer: N. Bone
Date: 1979 November 5/6
Location: Edinburgh. Latitude 55° 56′ 46.7″ N Longitude 3° 5′ 35.6″ W Altitude 30 metres
Instrument: 75-mm refractor, ×50 eyepiece

ZC Number	Mag.	Phase	Predicted Time (UT)	Observed Time (UT)	Estimated Error (s)	Comments
669	4.0	D	02^h 37.0^m	—		Lost in glare
671	3.6	D	02^h 46.9^m	—		Lost in glare
671	3.6	R	03^h 34.7^m	03^h 36^m 29.5^s	1.0	
669	4.0	R	03^h 43.1^m	—		Missed
677	4.8	R	04^h 51.0^m	04^h 51^m 8.5^s	0.5	Good timing
692	1.1	D	06^h 38.0^m	06^h 38^m 7.5s	1.0	Perfect sky
692	1.1	R	07^h 11.6^m	07^h 12^m 46.5^s	0.5	Good timing

Ultimately, results are analysed by the International Lunar Occultation Centre in Japan. The main result to emerge from detailed analysis of the accumulated data is the continual slowing of the Earth's rotation over time, which necessitates the addition, at irregular intervals, of leap seconds to Universal Time in order to keep pace with the time determined from highly accurate atomic clocks. The long-term slowing of the Earth's rotation is also discernible from careful study of pre-telescopic records of eclipses.

Sometimes, occultation observations can generate unexpected results. Most stars appear as tiny, unresolved point sources because they lie at enormous distances from us. Consequently, occultation disappearances or reappearances are usually instantaneous. One or two giant stars (Betelgeuse and Aldebaran are examples) can show appreciable disks when examined using specialised, professional interferometer equipment. Aldebaran has an apparent diameter of 0.03″, and may sometimes disappear slightly less than instantly. Other stars have been observed, unexpectedly, to fade gradually or step-wise during occultation disappearances, giving away their previously unsuspected double nature. There are now, however, probably few such close binaries remaining to be discovered.

Occultations of the planets do not in themselves yield much in the way of new scientific data, but remain popular observational targets for many amateur astronomers. Series of occultations of Saturn

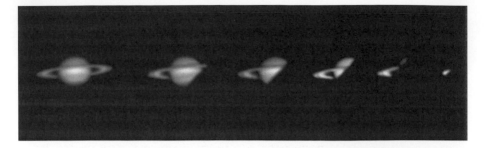

in the latter half of 1997 were widely observed and photographed, for example (Figure 5.3). Unlike the instantaneous disappearances and reappearances of stars, such events involving the planets take several seconds. The disappearances of Saturn in 1997, for example, took about a minute to become complete, as the Moon travelled gradually eastwards over the 40″ span of the planet's rings. Occultations of Jupiter's large disk (usually in excess of 40″) also take about a minute. Venus can sometimes show a very large disk, and is a striking sight when perched close to the lunar limb in the evening sky, as at its occultation in 1980.

Figure 5.3. The lunar occultation of Saturn on 1997 November 12, recorded by the Hampshire Astronomical Group's 610-mm reflector and CCD camera. Each exposure is 1/20 second. With the exception of the first two images, the interval between each pair is approximately 7 seconds.
Images by David Briggs

5.2.2 Grazing Occultations

Among the most interesting lunar occultation events of all are those that occur close to the northern or southern limb, so that the target star appears to pass almost tangentially to the Moon's disk. Careful observation of these *grazing occultations* can allow the limb profile to be mapped.

Unlike total occultations, grazes are visible only along narrow tracks a few kilometres wide, at most, on the Earth's surface. Observers too far to either side will see the Moon slip past without occulting the star, or a total occultation with disappearance and reappearance separated by a short interval. The most favourable grazes are those of bright stars in which the grazing limb of the Moon is dark, beyond the bright cusp.

Predictions for the ground tracks of grazing occultations are issued in annual yearbooks, allowing observing runs to be planned in advance. Revised predictions are sometimes issued a few weeks before the event. Graze observations demand teamwork, and provide excellent projects for local society groups. Sometimes, for major events such as a grazing occultation of a bright star such as Aldebaran, several societies may pool their resources and equipment to maximise coverage. Usually, observers will need to travel to the graze track to make the observation. It is a useful exercise for those organising the expedition to survey the observing site a week or two beforehand, and assess any likely problems. It might be necessary, for example, to obtain permission to observe from privately owned land, while local obstructions such as trees or hills will have to be avoided.

Graze observing requires much the same discipline as the recording of simple disappearance and reappearance events. The main differences are that several observers are required, simultaneously recording across the track, and that there may be multiple disappearances and reappearances over the course of several seconds as the target star is alternately hidden by mountains and revealed by valleys on the visible limb of the Moon. Each observer in the team needs to be alert so as to avoid missing short-duration events during the critical minutes of the Moon's appulse to the star.

Team members are spaced out along a line either perpendicular to, or at an angle across, the track. The records from each station represent a single slice, as it were, along the lunar limb. In combination, the record of disappearances and reappearances can be built up, as in Figure 5.4, to derive a limb profile. Some observers will see more events during the graze than others; those at the extreme limits of the track will see

Figure 5.4.
Theoretical lunar limb profile derived from a grazing occultation of Aldebaran on 1978 August 26, observed from the Scottish Highlands by four observers.
Courtesy Dr David Gavine

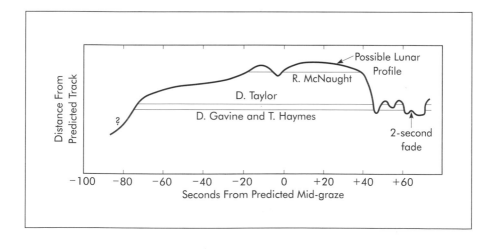

either a total occultation with no interruptions, or no occultation at all. Each observer's contribution is of equal importance. If sufficient observers are available, it is considered most reliable to have them work in pairs at the various stations across the track.

Observations will naturally be made with portable equipment. Ideally, members of an observing team will be experienced in occultation work, and may even bring with them the telescope which they use routinely at other times and with which they are therefore familiar.

Obviously, with the possibility of numerous disappearances and reappearances following one another in quick succession, the standard timing method outlined in Section 5.2.1 is not feasible: operating several stopwatches at once is simply too complex, and likely to lead to confused data. For graze occultations it is preferable instead to record an audio time signal on tape while the observer is calling out the sequence of events. One popular solution is to have a short-wave radio receiver tuned to a continuously broadcast time radio signal (MSF Rugby has been useful in the UK, while North American observers employ WWV), over which the observer will say "In" for disappearances and "Out" for reappearances. Subsequent playback and analysis allows the time and duration of each event to be determined.

Since observations will often have to be made in remote locations, tape recorders, radios and other equipment may have to function without access to mains electricity. Batteries are fine for operation, provided they are fully charged; spares must be taken to avert disastrous breakdowns during the observing run. Observations at isolated sites bring their unique problems. For example, observers covering the Aldebaran graze from the Scottish Highlands on 1978 August 26 had difficulty receiving the MSF Rugby time signal broadcast as a result of the local topography. This is exactly the kind of problem to be identified during the pre-event site assessment.

As with results from normal disappearance and reappearance events, reports should be sent either to the observers' national coordinating body (e.g. ALPO or the BAA) or to IOTA. Ultimately, the reports will be analysed by the ILOC in Japan to obtain profiles for the section of the lunar limb involved in the occultation. Such profiles, combined with data on the Moon's libration, are of value in forecasting the likely appearance of the lunar limb at total solar eclipses; the positions of valleys on the limb influence the duration of totality and the likely appearance of Baily's beads at third and fourth contact.

5.2.3 Electronic Observing

The advantages of time-stamped video recordings for meteor work have already been discussed (Section 2.1.4). Coupled to a reasonably sized driven telescope, low-light video systems offer at least an order-of-magnitude improvement in accuracy over visual observations for the timing of lunar occultations. Video recordings have the advantage that they are not subject to the physiological limitations of reaction time, and with frame rates of 25 per second, timings to within 0.04 second are possible. (The North American standard of 30 frames per second allows timings to an accuracy of 0.03 second.) For recording disappearance events, the advantages are obvious.

In addition to his work on recording meteors (Section 2.1.4), Andrew Elliott at Reading has applied his low-light video system to occultation timings. The camera is operated via a 200-mm catadioptric reflector, with time signals from an MSF receiver imported onto the recorded image as a running digital display (observers in the United States would use the WWV time signals).

Video recording also offers the possibility of more accurate recording of graze events. Some remarkable footage has been obtained showing stars blinking in and out of view behind the limb. As with audiotape recording equipment, there are problems of power supply to be overcome if video recording is to be employed on graze expeditions, but the results – occasionally spectacular, and usually certainly of scientific merit – justify the effort. Video has the obvious advantage of replay at leisure, and events recorded in this way are less likely to pass unnoticed, as they might in the heat of the moment during live observation!

While the electronic recording of occultations promises to bring about a leap forward in the accuracy of timings, the equipment remains expensive and comparatively specialised; the best results are obtained by experienced and dedicated observers. It will be many years yet before direct observation by eye and stopwatch is supplanted as the main method by which lunar occultations are recorded by amateur astronomers.

5.3 Lunar Eclipses

While total eclipses of the Sun are rare at any one location on the Earth, eclipses of the Moon can be observed without the need to travel. A lunar eclipse is visible from the whole hemisphere of the Earth for which the Moon is above the horizon.

In a lunar eclipse, the Moon passes through the dark shadow cone cast by the Earth into space opposite the Sun; lunar eclipses can therefore occur only at full Moon. Thanks to its orbital inclination (Section 5.2), the Moon is usually north or south of the shadow cone when full, and eclipses do not therefore take place each month: they occur only when the Moon is close to the ascending or descending node of its orbit at full. It is often the case that a lunar eclipse will precede or follow a solar eclipse (not necessarily visible at the observer's location, of course) by half a lunation.

The westward regression of the nodes of the Moon's orbit causes the "seasons" in which lunar eclipses can occur to slip gradually backwards. For example, the 1992 December lunar eclipse, seen against the stars of Taurus, has been followed by eclipses taking place earlier in the year and further west along the ecliptic, such as the 1996 September eclipse in Pisces.

Like solar eclipses, eclipses of the Moon may be partial or total, depending on how closely the Moon passes to the centre of the Earth's shadow cone. At the distance of the Moon's orbit, the umbra of the Earth's shadow has a diameter of about 9200 km, and the Moon can be completely immersed in it for a maximum of $1^h 47^m$ during a central passage. Most passages are above or below the centre, so that the duration of totality is less than this value. Obviously, during partial events not all of the Moon enters the shadow cone.

Since the Sun is an extended light source, the shadow cast by the Earth has two components: a central dark *umbra*, and a lighter surrounding region of partial shadow, the *penumbra*. During eclipses the Moon traverses the penumbra before and after entering the umbra. In some eclipses the Moon passes only through the penumbra, missing the umbra altogether. Such penumbral eclipses are far from impressive; indeed, the diminution of the full Moon's glare in such circumstances may not be detectable by eye. Obviously, umbral lunar eclipses are of more interest to the observer. At the very least, the brief "window" of darkness brought on by totality can be exploited for variable-star or other observations, which would otherwise be impossible around full Moon (Section 8.6.2). The eclipse itself, however, is also an interesting transient event which can be a rewarding visual and photographic target.

No two lunar eclipses are ever quite the same. Lunar eclipses vary in darkness from one to the next. Sometimes the eclipsed Moon appears quite bright and highly coloured, as on 1989 August 16/17 when, during totality, it was brownish red with a light yellowish border on the southern limb. A total eclipse on 1996 September 26/27 was also bright, again with a marked reddish coloration fringed in yellow on one limb. These colour effects result from the refraction of sunlight through the Earth's atmosphere at totality: because of the refractive properties of the atmosphere, the Earth's shadow is not completely dark. Red light travels most readily through the dense "ring" of air, just as it does through the thicker atmospheric wedge at sunset, so that the eclipsed Moon often appears reddish.

Sometimes, however, the atmosphere may have a heavier dust load, or greater amounts of cloud, so that less sunlight is refracted towards the eclipsed Moon. Very dark eclipses can then result. Most recently, the total lunar eclipse of 1992 December 9/10 was extremely dark. On this occasion the eclipsed Moon was a very dark grey, with little colour in evidence, and a very dim object once it was immersed in the umbra (Figure 5.5, *overleaf*). The 1992 December eclipse followed the volcanic eruption, over a year earlier, of Mount Pinatubo in the Philippines, which injected vast quantities of dust into the stratosphere, giving rise to spectacular twilight effects around the world (many will recall the purple twilights of 1992 October and November, for example).

The brightness of lunar eclipses is often described in terms of the Danjon scale, devised by the French astronomer André Danjon:

0   Very dark eclipse; Moon difficult to see, appears steely grey.

1   Dark; features such as maria or craters difficult to distinguish.

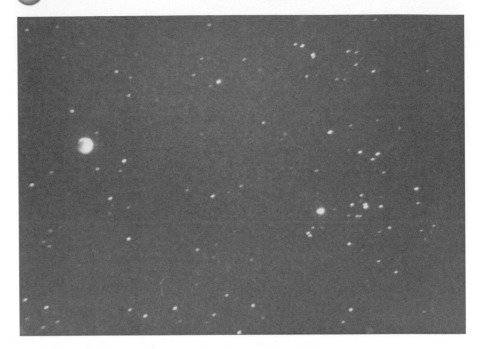

2   Dark red eclipse.
3   Eclipsed Moon quite bright, reddish. Maria easily
    visible.
4   Very bright eclipse, with Moon appearing coppery
    red-orange.

Bright eclipses, such as that of 1996 September,
would probably rate at 3–4 on the Danjon scale, while
the dark 1992 December event was more like 0–1.

Observed in binoculars or a small telescope, the
progress of a total lunar eclipse is interesting to follow.
As mentioned above, the penumbral stages are not
normally noticeable. Once first contact is made
between the Moon's leading limb (to the east on the
sky) and the umbra, a very obvious "nick" begins to
appear, growing rapidly over the next few minutes. To
begin with, there is little obvious reduction in the full
Moon's glare, but as time goes on the light level begins
to drop noticeably. About 30 minutes after first contact
the Moon will be about half-immersed and some gra-
dation may be apparent in the shadow. The curved
edge of the umbra is usually quite diffuse, and may
appear bluish, grey or yellow in contrast with the
coppery red hue of those parts of the Moon which are
more deeply immersed. During the very dark eclipse of

Figure 5.5. The dark,
totally eclipsed Moon
at 23:23 UT on 1992
December 9/10,
among the stars of
Taurus, captured using
a standard 50-mm lens
at f/1.8 and a
12-second exposure on
Kodak Ektachrome
400 film.
*Photograph by the
author*

1992 December, the edge of the umbra was markedly sharper than usual.

As totality approaches, the level of moonlight drops off more quickly – no longer casting shadows, for one thing. The Moon comes to resemble a somewhat distorted crescent, perhaps with lighter cusps or horns extending along the upper and lower limbs.

Finally, about an hour after first contact, the Moon becomes completely immersed. At this time of second contact, the trailing (westerly) limb is closer to the edge of the umbra, and may appear somewhat lighter. The limb nearest the edge of the umbra often remains brighter, sometimes with a yellowish or bluish fringe, compared with the rest of the eclipsed Moon during those eclipses when the Moon passes well above or below the centre of the shadow cone. For instance, during the 1996 April 3/4 total eclipse there was always a brighter "arc", starting on the western limb, then moving round the south of the eclipsed Moon and on to the eastern limb as totality progressed. At totality's end (third contact) sunlight reaches the leading (easterly) limb first, and the Moon re-emerges from the shadow to regain its full illumination over the next hour or so, with the immersion sequence in reverse.

During partial eclipses, of course, only some of the Moon enters the Earth's shadow. At the maximum extent of the eclipse, the partially eclipsed Moon appears as a crescent with perhaps a copper-tinted dark portion – much as it appears during immersion or emersion in a total eclipse.

Lunar eclipses can be recorded as sketches, indicating the extent and position of any colour fringes, for example, accompanied by notes of impressions made at the time. The times of first contact, start and end of totality, and re-emergence into sunlight can be recorded; sometimes these times are found to vary by several minutes from those given in the ephemeris. The umbra of the Earth's shadow shows some subtle variations in size, governed by the cloud and dust content of the atmosphere. Telescopic observations, in which the times of contact between the umbra and various craters on the Moon's disk, are collected by some organisations. These crater timings can be used to estimate the actual size of Earth's umbra (Karkoschka, 1996).

Lunar eclipses are ideal subjects for photography, particularly in colour. An undriven, tripod-mounted camera equipped with a standard 50-mm lens at $f/1.8$ may be used to photograph the eclipsed Moon and its stellar background, in exposures of 15–20 seconds' duration on ISO 400 colour film.

Telephoto lenses allow a closer view. A 135-mm $f/2.8$ telephoto will give reasonably large images. Exposures can be taken during the partial phase to record the progress of the eclipse, but these will have to be varied to compensate for the changing illumination. On ISO 400 colour film the early partial phases can be adequately recorded with an exposure of 1/30th of a second. Once half-immersed, the eclipsed Moon can be recorded with a 1/15th of a second exposure to show both the (overexposed) sunlit and eclipsed portions. Exposures need to be progressively longer as more of the Moon becomes immersed in shadow, and just before totality exposures of as much as half a second are required.

Undriven, tripod-mounted exposures of 3–5 seconds will record the colour of the totally eclipsed Moon, and should clearly show the contrast between its dark maria and lighter highland regions. It is interesting to compare identical exposures taken at different eclipses. The dark, greyish images of the 1992 December eclipse contrast with the bright orange eclipsed Moon of 1996 September, for example.

Table 5.2 lists lunar eclipses until 2003. The 2000 January 21 event will be excellent for early risers in western Europe, while the Moon will be high in the sky for those in North America, among the stars of Gemini. The 2001 January 9 event will be an evening spectacle for European observers, who will also enjoy the best views of the eclipse on 2003 November 8, with totality occurring when the Moon is high in the southern sky.

Table 5.2. Lunar eclipses 1999–2003. All times are UT

Date	First Contact	Second Contact	Third Contact	Fourth Contact	Maximum Partial Phase (%)
1999 July 28	$10^h 21^m$	—	—	$12^h 43^m$	39.9
2000 January 21	$03^h 04^m$	$04^h 06^m$	$05^h 23^m$	$06^h 26^m$	—
2000 July 16	$11^h 57^m$	$13^h 02^m$	$14^h 48^m$	$15^h 52^m$	—
2001 January 9	$18^h 41^m$	$19^h 49^m$	$20^h 48^m$	$21^h 56^m$	—
2001 July 5	$13^h 37^m$	—	—	$16^h 15^m$	48.7
2003 May 16	$02^h 04^m$	$03^h 15^m$	$04^h 06^m$	$05^h 17^m$	—
2003 November 8	$23^h 33^m$	$01^h 07^m$	$01^h 29^m$	$03^h 03^m$	—

References and Resources

ALPO Lunar Recorder: Julius L Benton Jr, 305 Surrey Road, Savannah, GA 31410, USA.

BAA Occultation Web Site: http://www.ast.cam.ac.uk/~baa/occalert.html

Boninsegna, R and Schwaenen, J, "Occultations" in *The Observer's Guide to Astronomy*, edited by P Martinez. Cambridge University Press (1994).

British Astronomical Association, *Guide for Observers of the Moon* (1979).

Doel, RE, "The lunar volcanism controversy". *Sky & Telescope* 92 4 26–30 (1996).

Dunham, DW, "Lunar occultation highlights for 1998". *Sky & Telescope* 95 1 96–9 (1998).

French, B, *The Moon Book*. Penguin (1977).

Hatfield, H, *Amateur Astronomer's Photographic Lunar Atlas*. Lutterworth (1968).

International Occultation Timing Association (IOTA), 2760 SW Jewell Ave, Topeka, KS 66611–1614, USA.

IOTA Lunar Occultation Web Site: http://www.sky.net/~robinson/ iotandx.htm

Karkoschka, E, "Earth's swollen shadow". *Sky & Telescope* 92 3 98–100 (1996).

Moore, P, "The Moon" in *Practical Amateur Astronomy*, edited by P Moore. Lutterworth (1974).

MSF Time Signals in the UK are broadcast at 2.5, 5.0 and 10.0 MHz.

Rükl, A, *Atlas of the Moon*. Hamlyn (1991).

WWV Time Signals in the United States are broadcast at 2.5, 5.0, 10.0, 15.0, 20.0 and 25.0 MHz.

Chapter 6

The Planets

The other major planets of our Solar System naturally attract a great deal of observational attention from the Earth's amateur astronomers. In particular, Mars, Jupiter and Saturn are rewarding observational targets. Mars's intriguing dark markings and bright polar caps are visible through a moderately large telescope at favourable apparitions. Jupiter's cloudscape presents a constantly changing appearance which can be followed in quite modest instruments. When tilted at their most wide-open aspect, the rings of Saturn are one of the most sublime sights in a telescope.

Of the other planets, innermost Mercury is always close to the Sun in our sky. For many observers it is an achievement simply to see Mercury during its brief forays at eastern (evening sky) or western (morning) elongation from the Sun; though it is occasionally to be seen in transit across the Sun's disk (Section 6.1.1). In the northern hemisphere, elongations are most favourably viewed on spring evenings and autumn mornings. At its brightest, Mercury is seen low in the gathering twilight as a pinkish or yellow mag. −1 to mag. 0 point of light. Telescopically, the planet's features consist of vague shadings – albedo (reflectivity) differences between different regions; at best, the average amateur observer might hope to detect the changing phase of Mercury and little more.

The two outer gas giant planets, Uranus and Neptune, are also comparatively unrewarding for the amateur observer. At a magnitude of +5.5, Uranus is, in principle, visible to the naked eye under good conditions. Certainly it can be picked up in 10×50 binoculars, and annual handbooks provide finder charts. Neptune is rather fainter, at mag. +7.7, but can again be found with binoculars provided a good chart is available. Neither of these worlds shows much in the way of detail in amateur-sized instruments. Uranus presents a tiny, featureless bluish disk of just under 4″ in diameter, while Neptune spans only 2.5″. Distant Pluto is an even less enticing prospect at mag. +15.

Mars, Jupiter and Saturn, however, are each of interest, and may be effectively observed with equipment similar to that recommended for lunar work (page 70). A 100-mm refractor or 150-mm reflector will give good views of Jupiter. Many would contend that a larger aperture is desirable, however, and certainly a reflector of 200 mm aperture or greater is preferable for observing Mars and the globe of Saturn. As in lunar work, atmospheric seeing (page 70) is a major constraint on the level of detail visible, and may also affect the visual appearance of features on the planets.

6.1 Venus

When an evening or morning "star" close to its maximum elongation of 47° from the Sun, Venus is unmistakable, shining at a brilliant magnitude –4, brighter than anything else in the night sky except the Moon. Venus is swathed in dense, reflective clouds which show little, if any, detail at visible wavelengths. Reports are occasionally made of subtle dark markings, present for only a day or two at a time, but these are difficult to see with any certainty. One problem with observation is simply that the planet is so brilliant that delicate albedo features in its cloud-tops may be swamped in the glare. It is therefore best to make observations in twilight before the sky is fully dark, or even during the daytime. Under good conditions – a clear, frosty winter afternoon, for example – Venus is readily visible to the naked eye in daylight. At other times it may be necessary to use setting circles on an equatorially mounted telescope to find Venus in the daytime; observers are strongly recommended to avoid the dangerous practice of sweeping for Venus in daylight with binoculars or a telescope, which carries the risk of unintentionally bringing the Sun into the field of view.

Observations in daylight certainly reduce the glare from the sunlit portion of Venus, and will clearly show the planet's phase. The elusive cloud features are, however, of low contrast, and may prove difficult to see in such conditions. Experienced observers recommend using a yellow Wratten 15 filter to help in reducing glare (Baum, 1995).

Its changing phase is Venus's most obvious feature. Observers have long been encouraged to make sketches, and estimate the phase from these. When close to superior conjunction, on the far side of the Sun from our terrestrial viewpoint, Venus shows an almost full or large gibbous phase. Around greatest elongation (east or west of the Sun) the phase is more akin to a half Moon; gauging the date at which exact half-phase (*dichotomy*) is reached is an important part of some observing programmes. Dichotomy usually comes earlier than theoretically expected at evening (eastern) elongations, and later at morning elongations – an anomaly described as the *Schröter effect* in recognition of its discovery by the German astronomer Johann Schröter in the 1790s. When Venus is near inferior conjunction, in line between the Sun and the Earth, it shows a crescent phase.

More obvious as transient phenomena in the Venusian atmosphere than the dusky shadings are occasional brightenings of the cusps of the crescent. From time to time one cusp may appear brighter than the other, perhaps as one polar cap is darkened by hazy material. Observers can record estimates of the intensity of Venus's cloud features on a six-point scale ranging from 0 (extremely bright) to 5 (exceptionally dark).

Short-lived anomalies on the terminator are sometimes reported. Most of the time the terminator of Venus is a sharp, regular curve. Irregularities in the form of bright extensions into the dark side, or dark indentations into the illuminated side, may result from temporary clouds at altitudes higher than the mean cloud-deck. Some of the reported features, however, are undoubtedly the results of glare and reflections in the telescope, and such observations have to be interpreted with caution.

Most enigmatic of all Venus's features is the *ashen light*, a general dusky greyish glow seen over the whole of the planet's visible night side when the phase is a narrow crescent. The ashen light was first seen by the Jesuit astronomer Giovanni Riccioli in 1643, and has been reported by countless observers since, including Schröter in the early 19th century. Many scientists

doubt that the ashen light is a real, intrinsic phenomenon in the atmosphere of Venus, considering it to
be an instrumental artefact. As with transient lunar
phenomena (Section 5.1), a mechanism to account for
the effect, if it is real, has to be sought. Given the
absence of a substantial magnetosphere in which electrons could be accelerated into the Venusian atmosphere, auroral processes can be ruled out as a possible
cause.

An *airglow* phenomenon, similar to that occurring
in the Earth's high atmosphere, is not impossible,
though the Venusian equivalent would have to be
much more intense than its terrestrial counterpart to
be visible over such a great distance. The weak terrestrial airglow arises when solar ultraviolet radiation
during the daytime causes electrons to be stripped
from atoms and molecules in the atmosphere, leaving
positively charged ions. Subsequently, during the
night, the electrons recombine with the ions, releasing
energy in the form of a weak background light which
ensures that, even far from artificial light pollution, the
Earth's night sky is never completely dark.

As with many other transient phenomena, the ashen
light and the short-lived dusky markings and terminator irregularities of Venus require further observation
and study. Most of the reports of these phenomena
continue to be made by amateur astronomers who
spend long hours observing the planet.

6.1.1 Transits of Mercury and Venus

As a result of the inclination of their orbits to the ecliptic (7.0° for Mercury, 3.4° for Venus), the two inner
planets usually reach inferior conjunction some way
either north or south of the Sun from our viewpoint.
On rare occasions, when inferior conjunction coincides
with the planet passing through either of the nodes
where its orbit intersects the ecliptic plane, Mercury or
Venus can be seen to *transit* the solar disk. Transits of
Mercury are more common than those of Venus.

Mercury can undergo transits at inferior conjunction in May (at its ascending node) or November (at
its descending node) following an evening elongation. Transits in November are twice as common as
those in May, for Mercury is near perihelion at

November transits, and near aphelion at May transits. During transits Mercury is seen as a tiny dark spot, with a diameter of about 11″, moving from east to west across the projected disk of the Sun (Section 4.1); the planet is too small to be seen with the (protected!) eye when in transit, so optical aid is essential for such observations. When transits occur at times of high solar activity, it is interesting to compare the intense black of Mercury with the penumbral and umbral regions of sunspots: Mercury is much darker, emphasising that sunspots appear dark only by contrast with the brilliance of the surrounding photosphere.

In the absence of an atmosphere around the planet, Mercury in transit appears as a reasonably sharply defined disk, depending on the seeing conditions. While there is little modern scientific value in doing so, observers may attempt to time the moments of first contact (when Mercury first makes a "nick" on the solar limb) and fourth contact (as it departs). In practice, these moments are difficult to determine reliably, especially if the Sun's limb appears to be "boiling" in the turbulence of a warm day. Slightly easier to time are second and third contact, when Mercury is tangential to the inside of the Sun's limb. Accurate timings of second and third contact are rendered difficult by the "black drop" effect, an optical illusion which gives the appearance that the planet – though fully silhouetted against the solar disk – is connected by a small extension to the Sun's limb.

Depending on the precise chord which it traces across the solar disk, a transit of Mercury can last for up to 9 hours. Most are shorter than this; the theoretical maximum duration would be observed only if the planet were to pass diametrically across the Sun's disk.

Transits of Mercury occasionally occur in pairs, roughly three years apart, the next such pair being on 2003 May 7 and 2006 November 8. A transit on 1999 November 15, with mid-event around 21:40 UT (and just days before a possible Leonid meteor storm – see Section 2.1.3) is favourable for observers in the Pacific region.

Transits of Venus occur in pairs separated by eight years, but with the pairs themselves separated by intervals of over a hundred years. The last pair were in 1874 and 1882. The next transits of Venus will be on 2004 June 7 and 2012 June 5, both at the descending node of the planet's orbit. The 2004 event will be seen to best advantage from the eastern hemisphere of the Earth, only the latter stages being favourable for observers in western Europe and North America. North American observers will see the beginning of the 2012 transit, which will be most favourable for those in the Pacific region. There is then a long gap until 2117 December 10 and 2125 December 8, on which dates Venus will be at its ascending node as it transits the Sun.

Unlike Mercury, Venus presents a sufficiently large disk (just over 1′) to be visible to the protected naked eye during transit. As with transits of Mercury, observers may wish to make timings of the limb contacts, but these are even more difficult to determine for Venus, which shows a very pronounced "black drop" effect, further complicated by the presence of a thick atmosphere through which sunlight can be refracted. Venus has been reported at past transits as showing, in the telescope, a faint partial halo-like extension beyond the solar limb as a result of the refraction of light in the planet's atmosphere. When Venus is fully in transit on the solar disk, its atmosphere shows as a diffuse ring surrounding the dark body of the planet; its silhouette is not as sharp as that of Mercury.

Historically, transits of Mercury and Venus were important in refining the distance scale of the Solar System. Simultaneous observations of transits from widely separated geographical locations allowed astronomers to estimate the solar parallax, a quantity from which the Earth–Sun distance (the astronomical unit) could be determined. Measurements of Solar System distances are nowadays more accurately obtained by radar measurements.

The first recorded transit of Venus was by Jeremiah Horrocks on 1639 December 4 (in the Gregorian Calendar), the second of a pair the first of which was in 1631; indeed, in 1631 there were transits of Mercury and Venus within a few weeks of each other – Mercury's transit on 1631 November 7 was the first to have been observed for the planet, by the French astronomer Pierre Gassendi at Paris.

6.2 Mars

Of the planets with orbits lying outside that of the Earth, Mars is surely the most celebrated and enigmatic, thanks largely to its association with the classic science fiction of H.G. Wells, Ray Bradbury and others.

Spacecraft exploration of the Red Planet has resolved some of the mysteries, but the widely reported detection of biological traces (held, in all honesty, by most biologists to be extremely doubtful) in a meteorite of Martian origin in 1996 has served to perpetuate the romantic public perception of Mars. Results from Mariner, Viking and subsequent spacecraft have shown quite comprehensively that Mars is an arid, dead world, though there is convincing evidence for the past existence of surface water, including a flash flood at the landing site of the 1997 Mars Pathfinder mission.

Through the telescope Mars appears as a yellowish-orange disk showing a bright, seasonal polar cap and grey dark markings. Like all the superior planets Mars is best seen at opposition, when it is directly opposite the Sun in the Earth's sky and at the same time closest in its orbit to us. Oppositions of Mars recur at intervals of approximately 26 months, and there is normally a period of favourable visibility for several months to either side of the opposition date.

Mars has a markedly elliptical orbit, so some oppositions are a great deal more favourable than others. When Mars comes to opposition close to perihelion, its disk has a comparatively large apparent diameter of about 25″, and will reveal detail even to quite small telescopes. At aphelion, however, Mars at opposition shows only a small disk, 14″ in diameter, and large instruments are required for productive observation. For observers in northerly latitudes the situation is made worse by the fact that perihelic oppositions occur when Mars is at a high southerly declination on the ecliptic, so that although the disk has as large an apparent diameter as possible, it has to be viewed low down, through a thick mass of unstable air, giving poor seeing conditions. Perihelic oppositions recur every 15 or 17 years, the next being in August 2003. Around perihelic opposition, the southern hemisphere of Mars is best presented towards the Earth; at aphelic opposition, we have a better, albeit more distant view of the planet's northern hemisphere.

The dark markings of Mars have been mapped by both professional and amateur astronomers for over three hundred years. Among the most prominent features is the triangular Syrtis Major, first depicted in a drawing made by Christiaan Huygens in 1659. Another notable feature is the light Thaumasia region, on the opposite hemisphere to Syrtis Major. A darker spot, Solis Lacus, at the centre of Thaumasia has led to the description of this region as "the eye of Mars" by some observers. Many of the other features are subtler, and long hours at the eyepiece are required for the observer to become familiar with them.

The long-term collection of observations by bodies such as ALPO and the BAA Mars Section has revealed gradual changes in the appearance of the markings. Some dark features alter in shape, or become more difficult to see as a result of albedo changes from one opposition to the next. These changes can be accounted for by variations in dust deposition: strong surface winds in the Martian atmosphere can raise and carry dust from the deserts, leading to the darker areas becoming covered from time to time. Features may also become darker, of course, if dust is removed. Following these long-term changes is a very rewarding pursuit for the dedicated observer (Figure 6.1).

Careful observation will also show short-term changes in Mars's appearance, resulting from local meteorological phenomena. These take the form of clouds or obscurations, appearing yellow or white. These phenomena, present on some nights but not others, were first noted when observers began to examine Mars more closely from the mid-1850s onwards. Yellow clouds are found near ground level, and are produced when dust is raised by near-surface winds. The white clouds are produced by the condensation of traces of water vapour (which comprises about 0.03% of the Martian atmosphere) at higher altitudes. White clouds imaged from the Martian

Figure 6.1.
Drawings of the onset of a regional dust storm on Mars by Dr Charles Capen, made using the 82-inch reflector at McDonald Observatory. (a) 1969 May 29, 06:00 UT (central meridian 310°); (b) 1969 May 31, 09:10 UT (central meridian 338°). The outline of the yellow-white dust cloud over Hellespontus is indicated by the sequence of dots and dashes. Between the two dates, the dust storm spread westwards. *Courtesy of Dr R.J. McKim and the BAA Mars Section*

a   b

surface by the Pathfinder probe in 1997 were reminiscent in appearance of the tenuous noctilucent clouds that form in the outermost fringes of the Earth's atmosphere (Section 3.2). The Martian atmosphere, which consists largely of carbon dioxide (CO_2), has a pressure at ground level of only 6 millibars – less than a hundredth of the Earth's atmospheric pressure at sea level.

With an axial tilt of 25° 11′ relative to its orbital plane, Mars, like the Earth, shows seasons. Summer in the southern hemisphere of Mars occurs when the planet is near its perihelion. The difference in axial orientation between the two planets means that Earth-based observers see the favourably presented hemisphere of Mars at opposition experiencing one season in advance of that in the terrestrial northern hemisphere. So, for example, at perihelic oppositions in August – in the Earth's northern-hemisphere summer – the Martian southern hemisphere is into its autumn. Mars's northern hemisphere is viewed at the beginning of its spring when at aphelic opposition in December or January.

The appearance of clouds on Mars is influenced by the seasons. Yellow (dust) clouds are most commonly observed when Mars reaches perihelic opposition, close to the time of its southern-hemisphere midsummer, for example. White clouds occur in a given hemisphere of Mars most often from spring to autumn.

Occasionally, as happened in 1956 and 1971, the number of outbreaks of yellow cloud increases, and the clouds can spread rapidly to become extensive. At 1971's perihelic opposition, particularly, the whole surface of Mars was obscured from view for several months by a planet-wide dust storm. More usually, the yellow clouds are confined to comparatively limited areas. The large, ancient impact basin of Hellas in the southern hemisphere, south of Syrtis Major, is a site from which many yellow clouds emerge, as is the nearby Mare Serpentis region, and also Solis Lacus and Chryse. These regions prone to yellow cloud activity are certainly well worth watching, especially at perihelic or near-perihelic oppositions.

Mars rotates on its axis once every $24^h 37^m$ (about 14.6° per hour). An observer at a given longitude on the Earth examining the visible disk of Mars at the same time each night will see the central meridian (the line of Martian longitude joining the planet's north and south poles) move by 10° eastwards – "backwards" – from one night to the next. Even with observing sessions of several hours' duration under favourable conditions, it may require a fortnight or longer for features which were on the averted side of Mars at the start of one observing run to come into view. For this reason the monitoring of Martian dust cloud activity has become an international project, involving amateur observers situated at different longitudes around the Earth. Alerts of possible yellow clouds or dust storms can nowadays be rapidly circulated by e-mail, allowing optimal coverage by those observers in favourable locations.

White clouds are more difficult to observe. Phenomena of the Martian summer, they are formed by the melting or sublimation of water from the polar cap, and are believed by some to consist of ice-crystal veils suspended in the high atmosphere. Lower-altitude haze or fog may also account for some white cloud observations: during the Martian spring, the polar cap is sometimes seen to be covered by a layer of haze. White clouds appear as spots or patches, mainly at equatorial or mid-temperate latitudes.

White clouds are often seen close to the morning or evening terminator of Mars. In the telescope, the leading or preceding limb of Mars marks the evening terminator, while the morning terminator is on the trailing or following limb. Mars is the only superior planet to show a marked phase, appearing somewhat gibbous when well away from opposition. Before opposition, the evening terminator appears *on* the visible disk within the preceding limb, while the morning terminator is presented inside the following limb after opposition. At such times white clouds may sometimes be seen extending beyond the terminator into the dark side, by virtue of their being – at high altitude – sunlit while the Martian surface below them is in darkness.

The high volcanic peaks in the planet's equatorial Tharsis bulge are sites where W-shaped *orographic clouds* often form as water vapour condenses from air masses forced to rise over them. Olympus Mons, Pavonis Mons, Ascraeus Mons and Arsia Mons (whose peaks all reach an elevation of 27 km above the Martian mean surface level) are all locations where orographic clouds appear, often in the afternoon (and, therefore, when they are seen on the preceding hemisphere of the observed disk).

Mars is best recorded by making annotated sketches at the telescope, and working them up as soon as possible after the observation into final versions. Since, as noted above, the planet rotates quite rapidly on its axis, it is important not to take too long to position the features on the sketch. The time of the observations should be recorded to within a few minutes' accuracy, and the longitude of the central meridian – which can be found from tables in annual publications such as the BAA *Handbook* – included in the final report.

Despite the advances brought about by the various spacecraft sent to Mars from the 1960s onwards, there is still much to be learned about Martian meteorology. Earth-based, amateur observations of the transient yellow and white clouds remain of value by adding to the long-term database, so leading to a better understanding of, for example, how the large-scale dust storms originate.

6.3 Jupiter

While Venus and Mars are both difficult and sometimes frustrating objects to observe with amateur telescopes, the largest of the planets, Jupiter, is a very different prospect. Even the smallest telescope will resolve Jupiter's disk, and show it to be crossed by horizontal dark belts. Amateur observers have studied Jupiter in detail for a very long time, and the value of the work of the BAA Jupiter Section and ALPO is very much recognised by professional planetary scientists (Beebe, 1996). As with the long-term record of meteorological phenomena on Mars, the amateur astronomers' databases provide the longest-running record of the continually changing appearance of Jupiter.

Jupiter takes nearly 12 years to orbit the Sun, returning to opposition at intervals of about 13 months, and appearing to move eastwards along the ecliptic by one zodiacal constellation from one year to the next. In the mid-1990s Jupiter was in the more southerly parts of the ecliptic, and less than ideally presented for observers in the Earth's northern hemisphere. Early in the 21st century, however, Jupiter at opposition will be high among the stars of Taurus and Gemini.

Viewed through a telescope suitable for lunar or planetary observing (150–200 mm in aperture), Jupiter presents an oblate disk with an equatorial diameter between 40″ and 50″. The polar diameter is somewhat less (about 37″ to 47″): Jupiter's oblateness is a result of its rapid rotation period of $9^h 50.5^m$ at the equator. It has a fairly circular orbit, to which its axis of rotation is tilted by only 3° 07′: unlike Mars, Jupiter is not subject to major seasonal variations in solar heating (insolation).

Jupiter's main, obvious features when viewed in a telescope are its dark *belts* and lighter *zones*, which have been recorded since the 1630s. These are more or less permanent, but are subject to localised changes from one night to the next. Most prominent are the North Equatorial Belt (NEB) and South Equatorial Belt (SEB), to either side of the Equatorial Zone (EZ). The SEB often appears double, the northerly and southerly components being designated the SEBn and SEBs respectively. Moving polewards from these, next in order come the Tropical Zones (NTrZ and STrZ), then the Temperate Belts (NTB, STB). Polewards of the Temperate Belts are the NTZ and STZ. At still higher latitudes, our oblique view restricts the reliability with which the various belts and zones can be resolved, and the convenient term "Polar Regions" is often used. Figure 6.2 shows a schematic representation of Jupiter's markings, as used to identify them in observational reports.

Jupiter's clouds, consisting in large part of ammonia compounds, are drawn out into the observed pattern of parallel bands and zones by extremely strong winds. In regions of wind shear and strong upwelling from the planet's deep gaseous interior, white spots – circulating weather systems – can form. Perhaps the most famous of Jupiter's features is the Great Red Spot (GRS), which lies in the STrZ, and cuts a "bay" or hollow into the SEBs.

Figure 6.2.
Schematic representation of the principal belts and zones of Jupiter, showing the standard nomenclature. SPR South Polar Regions; SSTZ South South Temperate Zone; SSTB South South Temperate Belt; STZ South Temperate Zone; STB South Temperate Belt; STrZ South Tropical Zone; SEB South Equatorial Belt; EZ Equatorial Zone. The northern belts and zones are similarly named; GRS is the Great Red Spot.

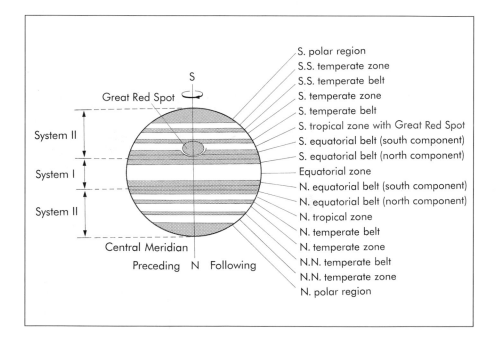

Visually, the GRS is often a disappointment to novice observers, appearing more as a pale, greyish ellipse (Figure 6.3); it has been more prominent and colourful in the past. The GRS is a site of atmospheric upwelling on a massive scale (it is considerably bigger than the Earth), and appears to be a more or less permanent feature, having been recorded since at least the 1660s.

Among other prominent long-duration features are three white oval spots in the STB, which have been present since the onset of a major disturbance in the STrZ in 1939. Each drifts gradually eastward around the planet, leading to encounters with the GRS at intervals of about three years.

Much of the value of amateur observations lies in recording the changing appearance of the Jovian cloud systems. This is best done, by making full-disk sketches or smaller-scale strip sketches of particular regions as

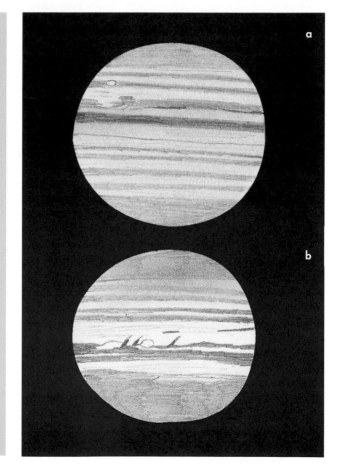

Figure 6.3. Two drawings of Jupiter made in 1978, using a 150-mm refractor at ×200. **a** 1978 January 30, 19:30 UT (central meridian of System I, 232.0°; of System II, 73.0°). The Great Red Spot is faintly visible at the preceding limb, with an STB white oval nearby. The SEBn is the darkest feature on the disk. **b** 1978 December 3, 00:48 UT (central meridian of System I, 87.0°; of System II, 130.0°). Now the NEBs is the darkest and most active region on the planet. *Drawings by the author*

they are carried into view by the planet's rapid rotation. At midwinter oppositions, when Jupiter may be above the horizon for 12 hours or more, the dedicated (and determined!) observer can follow the planet through an entire 360° rotation in longitude.

Details of wind-current systems can also be derived from *transit timings* of features such as spots, or features known as "spikes" and "festoons" on the edges of belts, as they cross the central meridian. At the time of the Voyager spacecraft encounters in 1979, amateur observers around the world contributed useful data to professional scientists via the International Jupiter Visual Telescopic Observation Project (IJVTOP). Observations collected through IJVTOP were used to identify features imaged in close-up by Voyagers 1 and 2. Similar ground-based observations were sought when Galileo's atmospheric probe dropped into the Jovian clouds in December 1996.

Transit timings are used to determine longitudes for features in Jupiter's atmosphere. Tables given in annual handbooks allow these longitudes to be calculated by the observer, whose timings should be accurate to within a minute or so. Jupiter's atmosphere does not rotate as a single solid body: the rotation period in the equatorial region bounded by the NEBs and SEBn is 9^h 50.5^m, compared with 9^h 55.7^m at higher latitudes. These different rotation periods, known respectively as System I and System II, are used to determine longitudes for features at equatorial and higher latitudes.

Jupiter's belts and zones show gradual variations from one apparition to the next, and the fine details of spikes, spots and other features may change over the course of a couple of nights. More rarely, there may be rapid changes in the major features over the course of a few weeks. At irregular intervals, for example, the SEBs may fade, following which its revival to prominence may be rapid, commencing as an outbreak of dark spot activity which spreads in both directions in longitude around the belt. A major fade of the entire SEB (SEBs and SEBn) between April and July in 1989 was followed by a reddening and darkening of the Great Red Spot. The EZ, and in the opposite hemisphere the NTB, also showed disturbances, indicative of upheaval on a global scale (Rogers, 1992). The SEB began to revive in the summer of 1990, probably before Jupiter had re-emerged from conjunction beyond the Sun. The reviving SEB was seen to contain dark sections, sometimes showing more pronounced brownish coloration than normal, interrupted by lighter "rifts". At the same time the GRS became fainter and less strongly coloured. The pattern was very similar to what was seen in previous SEB revivals, such as that of 1975. Following a secondary outbreak of activity in 1990 September, Jupiter settled down to a more familiar appearance.

6.3.1 The Shoemaker–Levy 9 Impacts of 1994 July

Among the most dramatic of all transient phenomena ever witnessed in the Solar System were the impacts with Jupiter of the shattered fragments of Comet Shoemaker–Levy 9 over the week of 1994 July 16–23. The sheer scale and energy of these impacts is exceeded only by cataclysmic solar flares.

Jupiter's deep gravitational well attracts many small bodies, both asteroids and comets, which pass close to it. There are well-established families of comets whose orbits have been modified by close passage to the giant planet, and several of Jupiter's smaller and more distant satellites are believed to be captured asteroids or cometary nuclei. Shoemaker–Levy 9 could also have become a long-term satellite, but for the close perijove (nearest approach to Jupiter) of its orbit. Following its discovery in 1993 March on a photographic plate taken at Mount Palomar by the team of Carolyn and Eugene Shoemaker and David Levy, the comet's orbital history was reconstructed. It became apparent that Shoemaker– Levy 9 (the team's ninth discovery, commonly referred to as SL-9) had been in orbit around Jupiter since at least 1929, with a period of about two years. At the perijove before discovery, gravitationally induced stresses had broken the nucleus of SL-9 into several fragments. Comet nuclei, consisting as they do largely of ice with a proportion of embedded dust, are mechanically weak, and several have been seen to break up during their passage through perihelion in the inner Solar System.

On the discovery plate, SL-9 appeared as an elongated "smear" – memorably described by Carolyn Shoemaker as a "squashed comet". High-resolution imaging with the Hubble Space Telescope showed the smear to consist of at least 21 individual small "nuclei", each with its own tail of gas and dust. Measurements soon showed the train of nuclei to be on a terminal orbit which would lead each to impact Jupiter's cloud-tops at around latitude 44°S over the course of a week or so.

Given over a year's advance warning of these events, both amateur and professional astronomers were well prepared to observe the results of the collisions. In the months leading up to 1994 July there was a great deal of speculation as to just what might be seen. Many were sceptical that the collisions would have any consequences observable from Earth. Others forecast that bright spots might be seen, and that the fireballs produced by the fragments' entry into Jupiter's atmosphere would temporarily brighten the Galilean satellites.

It was soon realised that the fragments of SL-9 would impact Jupiter just beyond the south-eastern limb as viewed from Earth, so that the entry events themselves would not be directly observable. The times of impact could be forecast with good accuracy, and many observatories were able to arrange their programmes accordingly. The SL-9 fragments were named along the train in order of their expected impact, from A to W. At the time of the impacts, Jupiter would be well past opposition and low in the south-western sky at the beginning of evening twilight. Observers at the latitudes of the British Isles and North America would have only a brief window of a couple of hours within which reasonable views of Jupiter could be obtained.

Fragment A hit the cloud-tops of Jupiter at 20:12 UT on the evening of 1994 July 16/17. An hour or so later, observers watching the giant planet through their telescopes were astonished to see a large, dark spot rotate into view over the eastern limb. Rather than producing visible flares or bright spots, the impacts gave rise to prominent dark scars. Over the course of the following week, Jupiter underwent what John Rogers, Director of the BAA Jupiter Section, described as "carpet bombing" along the latitude of 44°S. On average, the impacts were about seven hours apart, so that after a time it became difficult to distinguish which site was which. Figure 6.4 shows a strip-sketch of some of the impact sites.

Some observers reported brighter rings surrounding those dark spots seen soon after their formation. These rings were perhaps the effect of shock waves propagating outwards from the impact sites, and were gone within a few hours. Streaks of material were seen in other spots, some appearing darker to one side. The presence of such arcs of material was confirmed by observations from the Hubble Space Telescope, which from orbit had the clearest views, untroubled by atmospheric turbulence. The

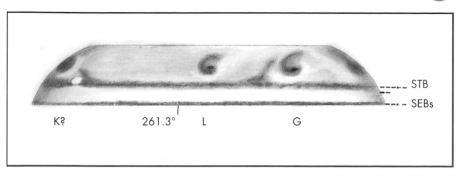

K?   261.3°   L   G

STB
SEBs

Figure 6.4. Strip sketch showing the effects of the impact of Shoemaker–Levy 9 on Jupiter, 1994 July 20, between 20:20 and 22:00 UT. A 216-mm reflector was used at magnifications ranging from ×216 to ×266 for the observations. *Drawing by Rob Bullen*

larger impact sites each showed a small, intense dark central spot marking the point of entry.

The largest fragments producing the greatest effects were G (July 18) and L (July 19), resulting in Earth-sized dark spots. The G impact occurred close to the site of the earlier (July 17) D impact, and the picture was further complicated when fragment S also entered Jupiter's atmosphere close to the same longitude on July 21. Seen together on the visible hemisphere of Jupiter, the G and L spots appeared uncannily like a pair of "eyes" in the telescope. Later impacts, following that of fragment L, appeared to have less dramatic effects. Some fragments, notably the small P1 and P2, and T, had no apparent effect on impact.

Intense scrutiny from professional observatories led to the detection of some of the immediate post-impact fireballs as plumes reaching up to 3000 km above the cloud-tops, and just visible over the limb of Jupiter. Material thrown up into these plumes condensed out and fell back onto the lighter clouds to produce the dark scars. Infrared observations showed the impact sites as glowing spots.

The impact spots faded and became less distinct within a few days of their formation, rapidly spreading out in longitude to form an "impact belt" around the planet. The spots' dispersal appears to have been a result of interactions with existing Jovian weather systems which, remarkably, seemed unaffected by the impacts. The resulting belt persisted until about July 1995, when it became fainter and broke up.

Debate continues over the original size (probably 1–10 km in diameter) of SL-9, and of the fragments produced on its break-up. Conservative estimates suggest that typical impacts (at 60 km s^{-1}) released energy equivalent to the explosion of 25 000 megatonnes of TNT. The consequences of a single such impact on Earth would be catastrophic!

The SL-9 impact events were unprecedented in modern observational history. Certainly, nothing like the observed dark spots had been recorded during the 150 years or so in which Jupiter has been under careful scrutiny. Some estimates suggest that Jupiter undergoes collisions with individual comets on this scale perhaps once every 2000 years. We were fortunate that the SL-9 impacts occurred at a time when appropriate observational technology – notably the Hubble Space Telescope – and expertise were available, allowing the events to be studied as fully as possible.

6.4 Saturn

One of the most spectacular sights in the telescope is Saturn at those times when its rings are presented in their full magnificence. Saturn, the outermost of the planets known to the ancients, lies at a mean distance of 9.5 AU from the Sun, almost twice as far as Jupiter. It takes 29.5 years to orbit the Sun, returning to opposition about a fortnight later from one year to the next. The orbit is inclined by 2° 29.5′ to the ecliptic plane. Of more importance to our view of the planet, and especially its rings, is Saturn's axial tilt of 26° 45′. When Saturn is at a high northerly declination on the ecliptic, we see the southern face of the rings opened out towards us, while the northern face is presented most favourably about 15 years later, when Saturn lies in the far south. At these times Saturn is at its brightest as a naked-eye object, thanks to the increased contribution from the reflective surface of the rings. When Saturn's thin rings (perhaps only a hundred metres deep) are seen edge-on, as they were most recently in 1995–96, the planet appears comparatively dim, at mag. +1.0 compared with a more prominent mag. −1.0 when the rings are opened at their widest.

While the rings are often seen as the main attraction, the globe of Saturn also repays scrutiny. Like Jupiter, Saturn is a gas giant, and the visible surface of the planet is comprised of its cloud-tops. Saturn has a rapid rotation (10^h 14^m at the equator) and an even more oblate disk than that of Jupiter: its globe has an apparent equatorial diameter of about 19.5″, compared with a polar diameter of 17.5″. Detail is harder to distinguish on such a small disk, and a telescope of at least 150 mm aperture, and preferably 200 mm, is recommended (Heath, 1995).

Not only is Saturn's disk much smaller than that of Jupiter, its features are also less distinct. In part, this is due to a deep haze layer overlying the cloud-tops, as found by the Voyager spacecraft in 1980 and 1981. Saturn does, however, have a system of belts and zones parallel to its equator. The belts sometimes appear reddish-brown, but are less often marked by spots than those of Jupiter.

The most prominent spots that arise on Saturn are short-lived white ovals in the Equatorial Zone or North Temperate Zone (the nomenclature for Saturn's belts and zones parallels that for Jupiter, illustrated in Figure 6.2). Particularly prominent examples were observed in 1876, 1903, 1933 and 1960. The discovery of a further great white spot in 1990 September lent support to the notion that the appearance of these spots follows an approximately 30-year cycle. It is tempting to suggest that the white spot outbreaks on Saturn are seasonal, given the planet's marked axial tilt. The greatest outbreaks tend to appear a few years after midsummer in Saturn's northern hemisphere, though not always at the same latitude on the planet.

Figure 6.5. The 1990 great white spot outbreak on Saturn, as seen through a 150-mm reflector at ×166 on 1990 October 10, 17:30 UT.
Drawing by David Graham, Director of the BAA Saturn Section

The 1990 outbreak (Figure 6.5) occurred in the Equatorial Zone, and was first seen as a bright white oval on September 24. Within three weeks it had spread (mainly eastwards, as seems to have been the case with previous white spots) around the EZ. By mid-October the spot had spread completely round the planet, giving rise to the appearance of a very light EZ. The disturbance had

begun to die down by 1990 December. Several lesser though significant white spots were seen in the following years, notably 1994 and 1996.

Like Jupiter, Saturn will be gaining in northerly declination in the opening years of the 21st century, and is likely to attract a great deal of casual attention as it climbs high among the stars of Taurus and Gemini. It seems unlikely, however, that another great white spot will be seen until around 2020. Indeed, there is strong observational evidence to suggest that such spots – as occurred in 1933 and 1990 – are part of a longer(55–60 year?) cycle, with outbreaks such as the 1960 event being relatively minor disturbances. A great white spot may therefore be essentially a once-in-a-lifetime occurrence for observers of Saturn, another being unlikely to appear until perhaps the 2050s (Heath and McKim, 1992).

6.5 Planetary Occultations

In the course of their passage around the ecliptic, the planets, like the Moon (Section 5.2), occasionally pass in front of stars, causing occultations. Since the planets' disks are very much smaller than that of the Moon, planetary occultations are much less common.

Before the advent of direct investigation by spacecraft, occultations of stars by the planets provided a means of examining the structure of planetary atmospheres. Occultations by Venus, Mars, Jupiter and Saturn indicated that they each had multilayered atmospheres, manifested as repeated flickering or fading of occulted stars; unlike occultations by the Moon, disappearances or reappearances at the limbs of these planets are gradual rather than instantaneous.

Among reasonably well-observed events in comparatively recent times was an occultation of the mag. +3.0 star Epsilon Geminorum by Mars on the night of 1976 April 7/8. Having come to a distant aphelic opposition the previous December, Mars presented a tiny disk just 6″ in diameter, and was relatively faint at mag. +1.1. Observers using larger instruments were able to follow the star until it was quite close to the limb of Mars. The disappearance took several seconds, during which the star flickered and faded as a result of the refraction of its light in Mars's atmosphere. Mars remained in front of Epsilon Geminorum for about 8 minutes.

Photometric observations by professional astronomers suggested a mid-occultation brightening of Mars, perhaps a result of "lensing" of the star's light by the Martian atmosphere; interestingly, a similar effect was seen and recorded during an occultation by Saturn's satellite Titan in 1989 (Section 7.3.1). About 18 hours before this star, 28 Sagittarii (mag. +5.8), was occulted by Titan, it was occulted by Saturn and its rings (Heath, 1992) as seen from the Americas. During the event, 28 Sgr was occulted first by the rings, then by the globe of the planet, and again by the rings following its reappearance from behind Saturn. At the time the north face of the rings was opened quite widely towards the Earth.

The multiple nature of Saturn's ring system has long been known. The three main components are designated A, B and C in order of decreasing distance from the globe. The Cassini Division is a major gap between the A and B rings; there is also the smaller Encke Division within the A ring. During the occultation on 1989 July 3, the outer A ring occulted 28 Sgr first, causing it to dim markedly within a few minutes of contact. Observers reported the star's light to fluctuate noticeably from minute to minute while seen through the A ring, and there were several interludes during which it disappeared altogether. The star became clearly visible again for 3–4 minutes as the Cassini Division passed in front of it, then it disappeared behind the B ring.

Disappearance behind Saturn's globe was not clearly visible since the C ring blocked the view for observers on the Earth. However, the reappearance was well recorded, about $1^h 4^m$ after Saturn had occulted the star. 28 Sgr reappeared gradually over the course of about 4 minutes. The star appeared in clear, dark space for about a further 5 minutes before the rings on the following side of the planet passed over it. As happened during the pass on Saturn's preceding side, marked fluctuations were seen, revealing structure (presumably the ringlets shown in close-up images obtained by the Voyager spacecraft) in the rings. The greatest dimming was again by the B ring. Photometric observations showed the star appearing to brighten while seen through the Cassini Division and, to a lesser degree, the Encke Division.

One of the more dramatic discoveries during a planetary occultation event was made on 1977 March 10, when professional astronomers followed the disappearance of a mag. +9 star (SAO 158687) in Libra behind Uranus. Uranus was expected to occult the star for about 25 minutes, and observers aboard the Kuiper Airborne Observatory – a converted aircraft – were monitoring the event, in part to obtain positional data to improve the targeting of the Voyager 2 spacecraft for its projected fly-by. In their photometric traces, they were surprised to record, some 40 minutes ahead of the planetary occultation, a 7-second disappearance of the star followed by a further four events over the next 9 minutes. The pattern was repeated, in reverse, after the occultation of the star by the planet, and Uranus was thus inferred to have a previously unsuspected ring system, whose existence was duly confirmed during the Voyager 2 fly-by in 1986 January.

A similar dark ring system exists around Neptune, though in this system much of the tenuous material appears to be gathered into arcs, rather than complete rings. The existence of these ring arcs had been suggested from occultation observations made as long ago as 1968, and re-interpreted in the light of the discovery of rings around Uranus. The arcs and other components of Neptune's thin ring system were imaged in more detail during Voyager 2's fly-by in 1989 August.

Occultations of stars by planets are rare events, but may often throw up much of interest. Accurate timings are made difficult by the usually large difference in magnitude between bright planet and faint star, often further compounded by the effects of poor seeing. Watching the brief flashes and fades as a star disappears behind a planet's limb, however, is as close as ground-based amateur observers are likely to get to probing the altitude structure of the atmospheres of our neighbouring worlds.

References and Resources

Association of Lunar and Planetary Observers (ALPO) Web Site: http://www.lpl.arizona.edu/alpo

BAA Planetary Sections Web Site: http://www.ast.cam.ac.uk/~baa/#SECTIONS

Baum, RM, "Mercury and Venus" in *The Observational Amateur Astronomer*, edited by P Moore. Springer (1995).

Beatty, JK and Levy, DH, "Crashes to ashes: A comet's demise". *Sky & Telescope* 90 4 18–26 (1995).

Beebe, R, *Jupiter the Giant Planet*. Smithsonian University Press (1996).

Heath, AW, "Saturn 1989". *Journal of the British Astronomical Association* 102 2 85–92 (1992).

Heath, AW and McKim, RJ, "Saturn 1990: The Great White Spot". *Journal of the British Astronomical Association* 102 4 210–19 (1992).

Heath, AW, "Saturn" in *The Observational Amateur Astronomer*, edited by P Moore. Springer (1995).

McKim, R, "The opposition of Mars, 1988". *Journal of the British Astronomical Association* 101 5 264–83 (1991).

McKim, R, "The dust storms of Mars". *Journal of the British Astronomical Association* 106 4 185–200 (1996).

Price, FW, *The Planet Observer's Handbook*. Cambridge University Press (1994).

Rogers, JH, "Jupiter in 1990–91". *Journal of the British Astronomical Association* 102 6 324–35 (1992).

Rogers, JH, "The comet collision with Jupiter: II. The visible scars". *Journal of the British Astronomical Association* 106 3 125–49 (1996).

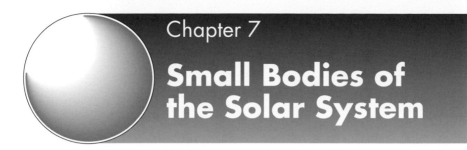

Chapter 7

Small Bodies of the Solar System

In addition to the Sun, Moon and major planets, the lesser members of the Solar System can also be participants in transient phenomena of interest to the astronomical observer. Meteor activity, produced by small particles – meteoroids – usually of cometary origin, has already been discussed (Chapter 2). Further small bodies of interest are the asteroids (or minor planets), planetary satellites and comets. Since asteroids and satellites are in general of a similar, rocky nature, it is reasonable to consider them in the same chapter. Cometary phenomena are rather different, and are discussed in the next chapter.

7.1 Asteroids

By mid-1998, some nine thousand asteroids had had their orbits characterised with reasonable certainty and been assigned a catalogue number. About as many again had been identified but remained to have their orbits properly defined, at which juncture they too will be assigned numbers (and, later, names) approved by the Minor Planet Center of the International Astronomical Union. While most asteroid discoveries are made with large, dedicated professional telescopes, amateur observers can still contribute. In the field of asteroid-hunting, Japanese amateur astronomers led by Tsutomu Seki have excelled. More recently the British observers Brian Manning and Stephen Laurie have made notable contributions.

To many novice observers it comes as a surprise to learn that as many as twenty asteroids may be visible, to a limit of about magnitude +9.0, in such humble equipment as 10 × 50 binoculars. In general, the lower-numbered asteroids (those discovered first) are the brightest. Asteroid (4) Vesta, particularly, is an easy object to pick out during favourable apparitions, when it may be as bright as mag. +5.0. Its brightness also makes Vesta a comparatively straightforward photographic target: undriven, a wide-field exposure on ISO 400 film with a standard 50-mm lens at f/1.8 will suffice to record it as a starlike point. Exposures taken a few nights apart will reveal its movement against the star background (Figure 7.1).

Some surprisingly high-numbered asteroids also come within binocular range under favourable conditions. For example, (324) Bamberga was fairly easily found at mag. +8.5 in 10 × 50 binoculars during 1991 October, while (532) Herculina can reach mag. +8.8.

The key to detecting asteroids with binoculars is to have a good chart, to a limit of mag. +9.5, and an ephemeris giving expected positions at intervals of a few days. Such ephemerides are provided for the brighter asteroids in annual publications including the RASC *Observer's Handbook* and BAA *Handbook*. Finder charts also appear from time to time in monthly magazines such as *Sky & Telescope* and *Astronomy Now*. A very useful standard atlas for this work is *Uranometria 2000.0*. For the brighter asteroids, detection usually

Figure 7.1.
Asteroid (4) Vesta (arrowed) moving against the stars of Cancer in 1988 January. **a** 1988 January 20, 23:22 UT; **b** 1988 January 22, 20:18 UT. Negative prints from 10-second exposures at f/1.8 with a 50-mm lens on Ektachrome 400 colour slide film.
Photographs by the author

entails "star hopping", via triangles, chains or other recognisable patterns of field stars, to the expected position of the interloper. If the observer becomes familiar with the field, movement of the asteroid relative to the fixed background should be obvious from one night to the next.

Hundreds of fainter asteroids come within the reach of even a small, 75-mm telescope. Again, skill (which comes quickly with practice) in using ephemerides and charts will be required.

Simply finding a given asteroid may be an end to itself for some observers; there are a few for whom the main challenge is to see the highest-numbered object possible. There is also some value in more systematic work. Visual magnitude estimates relative to background stars can be of some use. Several asteroids have irregular profiles, which lead to brightness variations as they rotate. For example, (15) Eunomia – another object well within binocular range – is markedly elongated, and may vary by up to half a magnitude over the course of its 6-hour rotation. Estimates made at intervals of 20–30 minutes throughout a night allow this variation to be followed. (4) Vesta has light and dark areas on its surface (as imaged by the Hubble Space Telescope in 1994) which lead to a 0.2-mag. variation in brightness as it rotates with a period of $5^h 20^m$. It is often sufficient to make estimates only once or twice per night.

Some asteroids may vary in overall brightness from one apparition to another, thanks to their markedly elliptical orbits. Like Mars, Vesta has perihelic and aphelic oppositions: at its 1989 opposition, for example, Vesta was on average half a magnitude brighter than at its opposition the previous year.

Visual magnitude estimates do have their limitations, however, and it must be said that close study of asteroids' rotation periods as derived from their magnitude variations really requires the use of a photoelectric photometer to provide results of high scientific accuracy.

7.1.1 Asteroid Occultations

An area where visual observers can contribute to scientific investigations of the asteroids is occultation work. Like the Moon and planets, asteroids do from time to time pass in front of background stars. The effect, as the asteroid closes on the star, is for the two images as viewed through the eyepiece to merge, giving an overall initial increase in apparent brightness. Then, as the asteroid obscures the star, the brightness dips abruptly for an interval of a few (perhaps 5), or maybe several (up to 30) seconds, before an equally abrupt brightening. By observing and timing these events, amateur astronomers can help in efforts to determine the profiles of asteroids, all but the largest of which are ellipsoidal or irregularly shaped.

In the first instance, an accurate timing by stopwatch of the duration of the fade allows the analyst to determine a chord across the asteroid's profile. Where several such observations are available across the occultation track, the chords can be combined to build up a picture of the asteroid's shape and size (Figure 7.2). The longest occultation will be seen where the chord cuts across the asteroid's major dimension (for most purposes, considered to be the major axis of an ellipsoid), while observers at the northern or southern extremity of the track will see only brief, grazing events. Negative events (i.e. no occultation seen) from either side of the track are equally valuable, as they help to place limits on the asteroid's diameter.

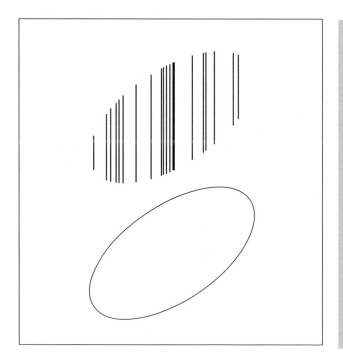

Figure 7.2.
Accurate timings of the duration of an occultation of a star by an asteroid can be used to derive chords across the asteroid's profile. These chords, in turn, allow a "best-fit" ellipsoid for the asteroid to be determined.

The most important observation is whether an occultation has occurred, and if so its duration, at a given location. Obviously the observing location should be noted, to the same accuracy as for lunar occultation reports (Section 5.2.1). If, in addition, the precise time of the event can be determined, so much the better. This can be achieved by recording the observation on tape, as for grazing lunar occultations (Section 5.2.2), with an audio time signal. There is also considerable scope for recording such events with a low-light video camera attached to an equatorially mounted, driven telescope.

A list of predicted events is essential for such work. Asteroid occultation predictions usually become available about 12 months in advance, and are issued by IOTA (Section 5.2.1) or the European Asteroid Occultation Network (EAON). Lists of possible favourable asteroid occultations are issued in paper form to regular potential observers, as for lunar events. More prominent events may be publicised in the popular magazines. IOTA and EAON also post listings at their respective Web sites. The Asteroids and Remote Planets Section of the BAA has an excellent Web site which provides up-to-date alerts for the next anticipated favourable event. Updates from this source also appear in the BAA *Circulars*.

Predictions give details of date and time for expected occultations, and of the asteroid and target star, and their respective magnitudes. Ideally, an event will give at least a 0.5-mag. drop in combined brightness, which is sufficiently obvious to allow a visual observer to react. In some cases the asteroid is much fainter than the star, and may not itself be directly visible during the observation, revealing its presence only when it causes the star to dim or disappear altogether. More detailed predictions provide an expected ground track for the occultation, and a chart of the field for the target star. Most predictions are for target stars brighter than mag. +10. Targets are usually quite faint, and are identified only by catalogue numbers.

The width of an asteroid's "shadow" – the ground track – on the Earth is virtually the same as its diameter. Since most asteroids have diameters of less than 200 km, an observer at a given location will be fortunate to be under the track of more than a couple of favourable occultations each year. Furthermore, there is sufficient uncertainty in the known motions of most asteroids that predicted tracks may be out by hundreds or even thousands of kilometres on the Earth's surface.

Predictions are subject to change as asteroids' apparent paths are recalculated on the basis of more recent observations. Indeed, amateur observers can often contribute work of great value by taking large-scale images of the fields of potential occulting asteroids in the weeks before the event for the purpose of astrometry. Reliable positions derived from astrometric-quality images can be used to refine predictions of both the times and likely tracks of asteroid occultations (Figure 7.3). It is certainly worth regularly checking the IOTA, EAON or BAA Web sites for up-to-date information in the run-up to attempted observations.

The equipment required is very much as for lunar occultations. Some of the brighter events can, at a push, be covered using binoculars, but the smaller field of view of a small telescope with an eyepiece magnification of ×50 may be preferable in offering fewer distractions! A portable instrument is an advantage since, as for grazing lunar occultations, it will often be necessary for the dedicated observer to travel to the expected track.

Observations should begin at least 15 minutes ahead of the predicted event, allowing plenty of time for the target star to be found (using the chart provided) and centred in the eyepiece. Main-belt asteroids (those residing between Mars and Jupiter, at a distance of between

Figure 7.3. An example of a ground track prediction for the occultation of a star by an asteroid, as issued via the BAA Web pages. Tracks are generated using *Asteroid Pro* software in conjunction with the *Guide 5.0* catalogue. *Courtesy of Dr Richard Miles*

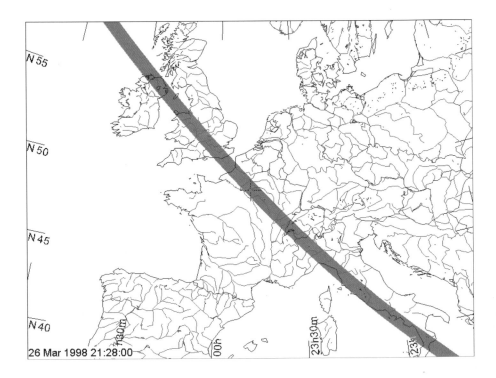

N 55

N 50

N 45

N 40

26 Mar 1998 21:28:00

2.0 and 3.5 AU from the Sun) move at about 30″ per hour, and if the prediction is reasonably accurate then asteroid and target should still be well separated at this point. It is not uncommon, however, for asteroid positions – particularly for faint, poorly observed objects – to be in error by a couple of arc seconds, which may in turn mean that the occultation occurs 5 or more minutes earlier or later than expected.

It is very demanding on observers, particularly in cold or windy conditions, to have to concentrate for as long as 30 minutes at the eyepiece. They will need to be at their most alert during the critical moment when the asteroid and star appear fused at the occultation itself. The approach of this moment can be anticipated for occultations by the brighter asteroids when both asteroid and star are visible. If the asteroid is too faint to be seen, however, there is little choice but to try to stay as alert as possible throughout the half-hour centred on the predicted occultation time. In many respects, this is where low-light video recording has a major advantage over visual observation at the eyepiece.

Observations made in the 15 minutes either side of the expected occultation may have additional value in revealing unexpected secondary events – perhaps of very short duration – resulting from small satellites in orbit around the main asteroid. Such observations are have now been made for a number of objects, including (6) Hebe, (18) Melpomene and (532) Herculina.

Asteroid occultation work demands not only alertness, but also patience. As a result of the inherent inaccuracies of the predictions, many events may turn out to be near-misses rather than occultations (though such negative results are still valid observations, and should of course be reported). Observers in the British Isles have had a considerable degree of misfortune thanks both to poor weather and off-track predictions. The first successful UK observation was finally made on 1996 November 9 by Richard Miles, who recorded the occultation of GSC 4695 545 by (892) Seeligeria. Observers elsewhere have enjoyed more success. Notably, in the United States excellent coverage of an occultation of the star 1 Vulpeculae (at mag. +4.60, an unusually bright target) by (2) Pallas was obtained by over a hundred observers on 1983 May 28. American observers have clocked up many further successes since.

So far, only three main-belt asteroids have been examined in relative close-up, two by the Galileo spacecraft when it was *en route* to Jupiter, and one by NEAR, the Near-Earth Asteroid Rendezvous probe, on its way to (433) Eros. In keeping with expectations from occultation work on other asteroids, (951) Gaspra, which Galileo flew by in 1991 October, and (243) Ida, imaged in 1994 August, proved to be irregular, heavily cratered bodies with sizes of $18 \times 11 \times 9$ km and $55 \times 24 \times 20$ km, respectively. Significantly, Ida was found to have a small (diameter 1 km) satellite, now named Dactyl, confirming that asteroids can have such companions, as first suggested by occultation observations. The images of (253) Mathilde obtained by NEAR in 1997 June showed that it too was heavily cratered, and $57 \times 52 \times 50$ km in size.

7.2 The Galilean Satellites of Jupiter

Jupiter's four major satellites – Io, Europa, Ganymede and Callisto in order of increasing distance from the planet – are familiar to anyone who has turned a telescope or even a pair of binoculars on the giant planet. Strung out in a line along the plane of Jupiter's equator, these Galilean satellites appear to shuttle back and forth from night to night. Close observation may often show one "overtaking" another over the course of a couple of hours, depending on their precise orbital configuration at the time. Annual handbooks and monthly magazines provide charts showing the continually changing order and alignment of the satellites, and are a valuable aid to identification. Many "desktop planetarium" computer programs also include ephemerides and simulations of Jupiter's four main moons.

The orbital periods of the Galilean satellites vary in accordance with Kepler's laws of planetary motion: innermost Io takes 1.8 days to complete one orbit, Europa 3.6 days, Ganymede 7.2 days and Callisto 16.7 days. Io never appears more than 2′ 17″ from Jupiter's limb. Callisto, on the other hand, can be as much as 10′ 13″ away (a third of the Moon's diameter), and it has been suggested that, at mag. +5.5, it should be visible to the naked eye provided that the observer can block off Jupiter's direct glare.

At their considerable distance from the Earth, none of the Galilean satellites really shows an appreciable disk in amateur telescopes. Ground-based observers have tried in the past to make out details, but even the Hubble Space Telescope can barely resolve anything. Close-up examination by Voyagers 1 and 2, and more recently by the Galileo spacecraft, has shown each of the Galilean satellites to be a fascinating word in its own right: Io with its sulphur-driven volcanism, Europa with a potential ice-locked ocean, Ganymede with its peculiar grooved terrain, and heavily cratered Callisto.

From the point of view of the amateur observer, the main fascination of Jupiter's principal satellites is in following their movements, including transits across the planet's visible disk, occultations behind its limb and eclipses in its shadow. The satellites themselves cast shadows, of course, which may also be seen to transit Jupiter's disk. Detailed knowledge of the Galilean satellites' movements allows forecasts of transits, occultations and eclipses to be made well in advance. Forecasts are published in annual handbooks, while the popular magazines will often highlight events of particular interest.

There is little scientific value to be gained by observing the various phenomena of Jupiter's satellites. In any case, the diffuse appearance and darkening of Jupiter's limb makes it extremely difficult to make accurate determinations of the moments of ingress and egress (the start and end, respectively) of transit phenomena. None the less, it can be an interesting pastime to watch these events in progress. During transits, the satellites and their shadows undergo ingress at the eastern (following) limb of Jupiter. The shadows appear as dark, slightly diffuse spots preceding the satellites before opposition, and following them after opposition.

The satellites themselves are not always easy to see against the disk, even through a sizeable (150-mm aperture, say) telescope. Io and Europa have particularly light, reflective surfaces and show low contrast with the planet's belts and zones. These two satellites may surprise observers unaware that a transit is in progress by becoming visible towards egress, as they are seen in better contrast against Jupiter's limb darkening. Ganymede and Callisto are easier to detect while in transit. These two outer Galilean satellites – Callisto in particular – have much lower albedos. Again in keeping with Kepler's laws, transits of Io take a maximum of about $1^h 20^m$, Europa $2^h 53^m$, Ganymede $3^h 40^m$, and Callisto $4^h 50^m$.

The Galilean satellites undergo occultation disappearances behind Jupiter's western (preceding) limb, reappearing at the eastern limb. The circumstances for observing occultations of the satellites are complicated, however, by the planet's shadow. Jupiter casts a long shadow cone into space to the west of the planet as we see it before opposition, and to the east after opposition. Eclipses of the satellites within the planet's shadow may result in one or other phase (disappearance or reappearance) of an occultation occurring when the satellite is, in any case, invisible. The effect of the shadow on whether an occultation event can be observed depends on the satellite, and on the proximity to opposition.

For Io and Europa, occultation disappearances before opposition occur when they are in eclipse. After opposition, Io and Europa reappear from behind Jupiter while still in the shadow, and become visible only when they re-emerge from eclipse some distance east of the planet's limb. Several months before opposition the more distant Ganymede and Callisto may be seen to undergo eclipse, then re-emerge briefly before occultation behind the planet. Closer to opposition, occultation disappearances occur for these satellites while they are in eclipse. Immediately after opposition, occultation disappearances for Ganymede and Callisto can be observed, but – as with Io and Europa – they do not become visible again until some time after reappearance, when they leave eclipse. Finally, well after opposition, towards the apparition's end Ganymede and Callisto can be seen to reappear from behind Jupiter's limb before becoming immersed in its shadow.

In the ephemerides for Jovian satellite phenomena published in annual handbooks, illustrations are usually provided for each month showing where, relative to the planet's limb, each satellite should undergo eclipse immersion and emersion. None of these events occurs very much farther than about 1′ from Jupiter's limb.

Of all the phenomena involving Jupiter's satellites, perhaps the easiest to observe are the eclipses. The satellites do not disappear immediately on entering Jupiter's shadow: immersion usually takes between 3 minutes (Io) and 9 minutes (Callisto), the satellites fading gradually from view. The events can be followed in even quite small instruments. The author once witnessed an eclipse of Ganymede, for example, using only a 110-mm reflector with an eyepiece that gave a magnification of ×30. During this event, which took several minutes from first fading to the apparent extinction of the satellite, Ganymede appeared to change colour from yellowish to steel blue before fading from view. Careful observation of these events may reveal further, subtle changes.

The three innermost satellites undergo transit, occultation and eclipse on each orbit. For three-year periods, centred on the time when Jupiter is at its farthest north or south of the ecliptic, Callisto can miss transit or occultation, appearing above or below the planet from our perspective. Thanks to its large orbital distance, Callisto will also pass north or south of the planet's shadow cone at these times and will therefore not undergo eclipses either.

7.2.1 Mutual Phenomena of Jupiter's Satellites

The four Galilean satellites have orbits close to the equatorial plane of Jupiter. Because of the slight inclination (3.1°) of the planet's orbit relative to this plane, there are two periods, each lasting a couple of months, in the course of Jupiter's 11.86-year orbit during which its equatorial plane is directly in line with the Sun. When this happens it is possible for the shadow of one satellite to fall across the face of another, giving rise to an eclipse. These two "seasons", about 5.9 years apart, occur when Jupiter is close to the ascending node (in Gemini) or descending node (in Sagittarius) of its orbit as seen from the Earth. Around these times the Earth also lies close to Jupiter's equatorial plane, and we may see occultations when one satellite appears to pass in front of another (Figure 7.4, *overleaf*). Multiple events, in which a satellite may be occulted and then eclipsed by the shadow of another (or vice versa), can also occur. Mutual occultation phenomena may be observed over a period of almost two years.

Observations of these events are of scientific interest, since they contribute to the refinement of the satellites' orbital parameters. Such data may be of value, for example, in targeting future spacecraft exploration of the Jovian satellite system. In practice, mutual phenomena involving the Galilean satellites are quite demanding to observe. A large telescope (of at least 150 mm aperture, and preferably larger) and high

1997 July 25 @23h41m18s UT

1997 July 25 @23h50m14s UT

©M.Gavin
C/serve 100772,47

WPO
Surrey-UK

magnification (×200 or more), and steady seeing conditions, are necessary if events are to be well observed.

Mutual occultations bear some similarities to occultations of stars by asteroids: the two satellites – usually fairly equal in magnitude – appear to merge and brighten, then the combined brightness of the pair drops during the occultation itself. The main scientific interest lies in obtaining the time at which the brightness minimum occurs. Where the disk of the occulting satellite fails to cover the target completely – a *partial* event – the drop in brightness may be too small to be noticed by eye; such events are not necessarily productive for visual observers, but require more specialised electronic photometry. Owing to the differing sizes of the Galilean satellites, combined with their varying distances along our line of sight, some occultations may be *annular*. Again, these may show only subtle light variations at mid-event. Under good seeing conditions, grazing events may take on the appearance of a single "fused' elliptical body at mid-occultation.

Figure 7.4.
Occultation of Europa by Ganymede, captured with a 300-mm catadioptric telescope and CCD camera on 1997 July 25.
Image by Maurice Gavin

Rather more favourable for the visual observer are eclipse events. These may be partial, total or even annular. Small partial eclipses may cause only a very small dimming of the target satellite. Total eclipses, however, will – like those that take place in Jupiter's much larger shadow – cause the target satellite to disappear from view. These events are generally of quite short duration, a matter of 5 minutes or less in most cases.

As for other events, the observer will need a list of predictions. These may be found in the popular literature (e.g. Westfall, 1991), and are also given in the BAA *Circulars*, or on the BAA Asteroids and Remote Planets Section Web page.

Some "seasons" for mutual phenomena are more favourable than others. For example, the alignments of Jupiter's equator with the Sun in 1979 and 1991 came close to conjunction at the end of the planet's apparition. On the other hand, the 1996–97 series of events was well placed, and many were seen when Jupiter was close to opposition. A further series of events will occur around 2002.

7.3 Saturn's Satellites

Like Jupiter, Saturn has an extensive satellite family of which at least four are readily visible in moderate-sized (150-mm aperture) amateur telescopes. Titan, at mag. +8.3, is the brightest; Rhea (mag. +9.7), Tethys (+10.2) and Dione (+10.4) are also relatively easy. When west of the planet, Iapetus can also reach mag. +10.2.

As a result of its orbital inclination (Section 6.4), there are two periods of about 9 months or so in each orbit during which Saturn's ring and equatorial plane lies close to that of the Earth's orbit. Around these times – when Saturn is in either Pisces or Virgo – we see the ring system edge-on. The satellites' orbits also appear (like those of Jupiter's Galilean satellites) more or less edge-on, and mutual occultations and eclipses may be seen. The satellites and their shadows may also undergo transit across Saturn's disk, but these events are much more difficult to observe than those involving the Galilean satellites. Only Titan's shadow transits are really a feasible target for amateur telescopes, and then only in larger instruments under excellent seeing conditions. Close to the time when Saturn's equatorial plane is in line with the Sun, its satellites can also undergo eclipse in the planet's shadow. Titan can be immersed for about 2 hours, Rhea for about 3 hours. The last series of events took place during the ring-plane crossing of 1995–96 (Nicholson, 1995).

7.3.1 The Occultation of 28 Sagittarii by Titan

A rather more dramatic event, involving Saturn's major satellite Titan, occurred on the night of 1989 July 3/4. Some months previously predictions had been issued by Lawrence H. Wasserman, of Lowell Observatory, suggesting that Titan would occult the star 28 Sagittarii around 22:42 UT on this night, about 18 hours after the occultation of the same star by Saturn and its rings (Section 6.5). The initial projected ground track for Titan's "shadow" gave a very favourable forecast for European observers, with the southern limit at the Mediterranean.

A revised prediction issued closer to the event was less hopeful for the British Isles, the track being shifted southwards, so that only the extreme south of the UK *might* see a grazing event. The night of 1989 July 3/4 proved clear (for once!) across much of the British Isles and, indeed, north-west Europe. A great many observers set up their instruments and watched in hope to see whether Titan *would* occult 28 Sgr from their location, bearing in mind the intrinsic uncertainties in the predictions for such events.

The event was similar, in some respects, to an asteroid occultation, with the critical difference that Titan has a dense atmosphere. At mag. +5.8, the target star was over two magnitudes brighter than Titan (mag. +8.2), so any occultation would result in a major drop in brightness. Visual observers began watching the field closely some time ahead of the possible event. Up to about 30 minutes beforehand, Titan and 28 Sgr could be resolved in a small telescope at ×75 magnification. Continuous monitoring for many began at $22^h 25^m$ UT, about 15 minutes ahead of the predicted occultation time, by which time the two images had already fused. The combined image of Titan and 28 Sgr appeared yellowish to the eye.

Observing from southern England, the author saw the occultation begin at $22^h 39^m 40^s$ UT, immersion starting with a flicker followed by a gradual fade over

10 seconds. Across north-west Europe, other observers witnessed much the same: contrary to the revised prediction, Europe was well under the track, and the British Isles close to the central line. The occultation lasted until 22^h 44^m 50^s UT – for 310 seconds – in southern England, with the return to maximum light being, again, gradual. During the occultation the combined image of Titan and 28 Sgr changed from its earlier yellowish appearance to a dull bluish.

The most remarkable feature of the event was a central "flash", recorded at mid-occultation by a number of visual observers (Figure 7.5). Around 22^h 42^m 09^s UT, the merged image of Titan and 28 Sgr brightened slightly for a couple of seconds, before abruptly dimming again. Valuable records of the timing of this flash were obtained by those using audiotape to record their observations.

The central flash – and, indeed, flickering during immersion and emersion – were confirmed by video recordings made at the eyepiece by several observers, and by the results of photoelectric photometry carried out by others. The flash is believed to have resulted from refraction of the star's light by Titan's dense atmosphere, and was apparently most prominent for those in the south of England, closest to the centre of the occultation track. It was also noted, however, as far away as St Andrews, 650 km to the north.

Figure 7.5.

Photometric record of the occultation of 28 Sgr by Titan on 1989 July 3. The rapid fluctuations in brightness at immersion and emersion are readily apparent, as is the central "flash", believed to have resulted from refraction of the star's light by Titan's atmosphere. *Reproduced by kind permission of Dr Richard Miles*

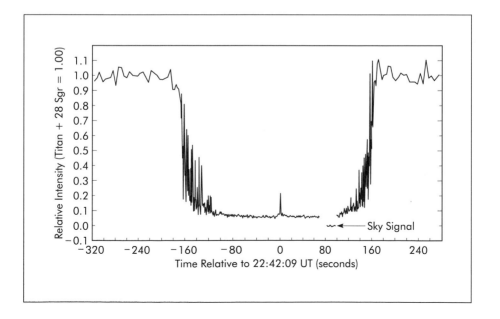

This remarkable, rare occultation was seen across much of Europe, from Norway and Sweden in the north, through the British Isles and France, and down to the Canary Islands and Malta.

References and Resources

BAA Asteroids and Remote Planets Section Web Site: http://www.ast.cam.ac.uk/~baa/occalert.html

European Asteroid Occultation Network Web Site: http://www.xcom. it/cana/EAON

Kowal, CT, *Asteroids: Their Nature and Utilization*. Wiley-Praxis (1996).

Miles, R and Hollis, AJ, "The occultation of 28 Sagittarii by Titan". *Journal of the British Astronomical Association* 104 2 61–76 (1994).

Minor Planet Center Web Site: http://cfa-www.harvard.edu/sfa/ps/ mpc.html

Nicholson, PD, "Saturn again turns ringless". *Sky & Telescope* 90 2 72–6 (1995).

Westfall, J, "Watch Jupiter's moons play tag". *Astronomy* 19 1 58–63 (1991).

Chapter 8

Comets

Among the small bodies of the Solar System, comets attract perhaps the most attention from amateur observers. In a good year perhaps fifteen to twenty comets may come within range of amateur telescopes, maybe a couple of which might even be sufficiently bright to be seen with the aid of 10×50 binoculars. Most are rather faint, and require skill and patience if they are to be observed. Certainly, few come to resemble the popular image, with a long tail streaming away from the head: the objects followed by regular, specialist comet observers are more usually faint, fuzzy patches with little or no obvious tail. Only rarely are we treated to spectacular objects like C/1996 B2 Hyakutake, which appeared in the spring of 1996, or C/1995 O1 Hale–Bopp, which reached naked-eye visibility a year later. The arrival of such bright objects is unpredictable, and they may be easily visible for only a few weeks.

(The International Astronomical Union standards for comet designations adopted in 1995 are as follows. Long-period comets, nominally those with periods longer than 200 years, are given the prefix C/, coupled with the year of discovery and a letter indicating the fortnight, and order in that fortnight, of discovery, as in the examples above. Short-period comets, those with periods less than 200 years, are assigned the prefix P/ and a number indicating the order in which the periodicity was established, as in 1P/Halley. Defunct comets are prefixed by D/.)

8.1 The Nature and Origin of Comets

The observed phenomena associated with comets can be understood in terms of the now widely accepted "dirty snowball" model, first put forward by Fred Whipple in 1949. In Whipple's model the main mass of a comet is concentrated in a small nucleus, perhaps only a few kilometres in diameter. This nucleus consists of water ice, and ices of gases such as carbon monoxide (CO), with an admixture of dusty material. At great distances from the Sun such a nucleus is inert, basically indistinguishable from an asteroid in appearance.

Should the nucleus come closer to the Sun than the orbit of Jupiter (5.2 AU), heating by solar radiation will cause some of the gas to *sublime* (passing straight from the frozen, solid state into the gaseous state) forming around it a thin temporary "atmosphere", the *coma*. Fluorescence emission by ionised CO in the coma gives it a bluish appearance. As the coma grows, it is also affected by the solar wind flowing outwards from the Sun. Many comets have a teardrop-shaped coma, rounded in the direction towards the Sun and coming to a point "downwind". With increasing solar heating, the coma will continue to grow and dust will be released from the nucleus, adding to the coma's brightness.

Material dragged downwind from the coma may start to form a more or less straight, bluish *ion tail* or *gas tail*. Released material with a larger particle size begins to fall away on parabolic trajectories behind the comet as a more curved *dust tail*. The release of larger particles, some perhaps as much as a centimetre in diameter, may, over successive returns of the comet to perihelion in the inner Solar System, lead to the development of a *meteor stream* (Chapter 2).

So far, only one cometary nucleus has actually been examined in close-up – that of 1P/Halley on the night of 1986 March 13/14 by the ESA Giotto spacecraft. The images returned by Giotto suggest that Whipple's "dirty snowball" model is essentially correct, at least for Halley's nucleus. The nucleus of Halley's Comet is an irregular body, some 15×8 km in size. The comet was seen to have a dark crust, and appeared to have a number of separate active regions from which jets of gas were emerging, carrying dusty material with them.

Comets are primitive bodies, remnants from the formation of the Solar System. It is believed, from studies of the dynamics of how the Solar System is likely to have formed, that comet nuclei formed in a region about the same distance from the Sun as the orbit of Uranus. From here they were flung out by gravitational perturbations into a vast halo surrounding the Solar System. The existence of such a shell of cometary nuclei was first proposed by the Dutch astronomer Jan Oort in 1950, after whom it is now known as the Oort Cloud.

The Oort Cloud extends to about one-third of the way to the nearest star, and estimates suggest that there may be as many as 10^{12} cometary nuclei resident in it. Oort's original suggestion of the existence of such a cloud was based on observations of new long-period comets, which appeared to enter the inner Solar System from completely random directions. More recent theoretical modelling has suggested that the Oort Cloud has extensions inwards towards the Edgeworth–Kuiper Belt, which is believed to constitute a reservoir of cometary nuclei closer to the ecliptic plane at distances from 30 AU (Neptune's orbit) out to about 1000 AU.

The comets we observe may have undergone a complicated orbital history. Comet nuclei resident in the Oort Cloud will stay there until gravitationally perturbed, perhaps by the passage of another star relatively close to the Solar System as the Sun orbits the Galaxy. Comets falling inwards from the Oort Cloud may then spend some time in the Edgeworth–Kuiper Belt before being further perturbed into their first passages through the inner Solar System.

The Edgeworth–Kuiper Belt is a source of new, long-period comets, which may take thousands or tens of thousands of years to complete their long, highly elliptical orbits. Close passages to the major planets – Jupiter in particular – may alter the orbit of such a long-period comet so that it makes more frequent returns to the inner Solar System, becoming a short-period comet. Halley's Comet is regarded as the archetypal short-period comet, returning to perihelion every 75 to 76 years. Others, such as 2P/Encke (3.3 years) have even shorter periods; several, following repeated close encounters with Jupiter, have periods close to 6 years.

The observational characteristics of a comet depend on its nature. For example, the most spectacular are believed to be those newly arrived in the inner Solar System from the Oort Cloud or Edgeworth–Kuiper

Belt, as on their first visit to the warmer neighbourhood of the Sun they have more volatile material to release. However, some first-timers – such as the infamous C/1973 C1 Kohoutek – turn out to be less impressive than expected, perhaps because their (postulated) higher initial rate of dust emission chokes off the active vents on the surface of the nucleus. The spectacular C/1995 O1 Hale–Bopp is believed to have undergone at least a few previous perihelion passages, as a result of which its nucleus was more uniformly active than it would have been on a first passage. Periodic comets which make frequent returns to perihelion become progressively depleted in volatile material and dust, and put on less impressive displays. 2P/Encke, for example, is now a fairly unimpressive object. Several other short-period comets are faint.

Eventually the nucleus may become completely inactive, and the comet comes to resemble an asteroid throughout its orbit, with no coma produced near perihelion. An example of such an object may be 107P/Wilson–Harrington, which showed cometary activity, including a tail, when found in 1949 but now appears asteroidal.

In general, then, the most impressive comets for amateur observers tend to be new discoveries, while periodic, predictable objects often prove to be faint, difficult targets.

8.2 Comet Discoveries

The movements of short-period comets, if sufficiently well known, can be used to forecast the likely time and visibility of their next returns. Each year the BAA *Handbook* lists orbital elements and, in more favourable cases, detailed ephemerides for about fifteen objects expected to return to perihelion (or at least to become visible again) that year. Many "desktop planetarium" computer programs are capable of generating detailed finder charts for comets, provided that their orbital elements are available. Positions from the BAA *Handbook* can also be plotted on a good atlas (*Uranometria 2000.0* is a well-established standard for comet work), allowing the observer to attempt the recovery of known comets, even when they are faint.

The discovery of new comets (which, by their very nature, cannot be forecast), is a much more difficult process. While occasional serendipitous finds are made, comet-hunting is strictly a pursuit for the extremely dedicated. Perhaps the most productive approach is systematic visual *sweeping* of the sky with a pair of large-aperture binoculars (15×80, for example). For such searches to be successful, however, the observer will need to spend a great many (several hundred!) hours becoming familiar with the night sky. There are many faint, fuzzy objects out there which are *not* comets – as Charles Messier knew well when he began to compile his famous catalogue (published 1771–81) of what we now refer to as deep-sky objects: galaxies, nebulae, globular clusters *et al.*

Messier, using a small and quite primitive telescope, had enough problems. The modern comet-hunter, with much better-quality optics, will pick up not just the brighter objects catalogued by Messier but also hundreds of fainter galaxies and nebulae. This is why a thorough working knowledge of the sky is required. *Uranometria 2000.0* will help in identifying the various NGC, IC and other non-cometary objects which will, inevitably, be swept up and not recognised in the first few viewing sessions.

Since comets are usually at their brightest when close to the Sun, searches should be concentrated in the parts of the sky adjacent to, or even in, the morning and evening twilight. Searching should begin in the evening as soon as the sky is reasonably dark, and in the morning should continue until the light of dawn becomes too bright, at times when moonlight is not a nuisance.

The recommended sweeping pattern is parallel to the horizon, working upwards from the light western horizon in the evening, and downwards to the eastern horizon in the pre-dawn. Sweeps should be slow, and a tripod mount will be essential for a pair of heavy binoculars. Suspected objects must be checked against the atlas and, ideally, for movement relative to the background if time permits. Observers must exercise extreme caution against raising false alarms. Internal reflections in the binoculars' optics – "ghost" images of bright stars at the field's periphery – have been the source of some spurious claims. Experienced observers keep in touch with one another, so that suspects can be quickly confirmed independently before the authorities (the IAU, for example) are contacted. It is certainly preferable to verify any potential discovery through an intermediary before contacting the professionals. *The Astronomer* magazine's observer network carries out excellent work in this capacity for comets (and also for novae and supernovae – Chapter 10) and active participation in its work is strongly recommended to anyone thinking of embarking on a search programme.

Only a few dedicated observers have taken up the challenge of comet-hunting in the late 20th century, but some notable successes have been clocked up by those amateurs who do this work systematically. Among the leading comet discoverers of recent decades have been the Japanese observer Minoru Honda, who discovered twelve between 1940 and 1968, and the Australian Bill Bradfield with seventeen between 1972 and 1995. In the United Kingdom the leading comet-hunter of recent times has been George Alcock, with five discoveries between 1959 and 1983. Don Machholz and David Levy have made several important discoveries from North America.

Even with professional sky surveys, and the efforts of the handful of existing comet-hunters, poor weather or other circumstances at their observing sites may mean that the next bright comet is discovered by a relative newcomer to the field.

8.3 Observing Comets

Success in observing comets depends on a number of factors, not least of them a dark observing site, preferably with clear views to the eastern and western horizons – where those objects close to the Sun will most often appear. As with observing deep-sky objects, the absence of artificial light pollution is desirable, though ever more difficult to achieve.

8.3.1 Comet Ephemerides

In order to find a comet in the first place, the observer has to know where to look. As mentioned above, expected positions are given in the BAA *Handbook* for known periodic comets. However, some periodic comets may "misbehave", appearing well off their expected tracks. Extreme examples are 109P/Swift–Tuttle – the parent of the Perseid meteor stream (Section 2.2) – which was recovered in the autumn of 1992, over 10 years later than some had predicted, and 23P/Brorsen–Metcalf, which made a late return in 1989 August. These objects were both subject to considerable *non-gravitational perturbations* which slowed them in their orbits by the "rocket effect" of jets emerging from their nuclei. More recently, many observers had difficulty in locating 55P/Tempel–Tuttle – the parent of the Leonids (Section 2.1) – in late January 1998 until a new, revised ephemeris became available. Updated ephemerides can be found on the BAA Comet Section's Web page, while the NASA Jet Propulsion Laboratory also maintains a useful site with up-to-date information on current comets.

For newly discovered comets, speed of notification may be of the essence. Postal sources, such as the BAA *Circulars*, have been valuable in the past, and are still useful for providing updates during a comet's apparition if time permits. More rapid dissemination of news over the Internet or by electronic mail would appear to be the future preference. Speedy notification is particularly vital for comets likely to be visible for only a short time. For instance, IRAS–Araki–Alcock (Figure 8.1) was visible from northerly latitudes only for a week or so early in May 1983, and many observers were lucky to catch a glimpse of it as it passed rapidly by; delays in issuing the news of the comet's appearance would have prevented many from seeing it at all.

Figure 8.1. C/1983 H1 IRAS–Araki–Alcock was the brightest comet of the 1980s, appearing as a mag. +2 object in 1983 May. This 40-second exposure was made on Ilford HP5 ISO 400 black-and-white film, using a 50-mm f/1.8 lens, on 1983 May 9 at 23:07 UT.
Photograph by the author

Even with knowledge of a comet's expected position, all is not necessarily plain sailing. The ephemerides often give a predicted magnitude based on earlier behaviour of the comet. This is often a total, *integrated* magnitude. As many found with 55P/Tempel–Tuttle early in 1998, however, an integrated magnitude of +8.0 can count for very little if the comet's light is spread over a wide area of sky, perhaps 0.5° or more across. Diffuse comets will require excellent-quality clear skies (no haze whatsoever) and a dark observing location. Furthermore, observers will need to allow time for their eyes to become *dark-adapted*, undergoing a physiological change to night vision which takes about 20 minutes and allows optimal detection of faint objects. Extraneous artificial lights can, of course, ruin this dark adaption in moments.

A large, faint diffuse comet has low contrast with the sky and the observer will therefore need to search the expected field carefully. As with finding asteroids (Section 7.1), a routine of star hopping, aided by a reasonably detailed atlas, is usually essential. It should also be borne in mind that comets move against the star background from night to night, and it will therefore often be necessary to interpolate between two sets of ephemeris coordinates.

8.3.2 Comet Magnitudes

Two of the fundamental observational properties of a comet, then, are its brightness and how diffuse or condensed it is. The former is influenced to an extent by the latter. The eye will have less difficulty at the eyepiece in detecting a condensed object. In reporting the appearance of comets, observers are encouraged to give an estimate of the *degree of condensation* (DC), ranging from 0 (extremely diffuse) to 9 (strongly condensed or nearly starlike).

Magnitude estimates are useful, especially if large numbers from many observers are collected over a long span of a comet's apparition. Corrected for the varying Sun–comet distance to derive an absolute heliocentric magnitude, these estimates can be compiled into a light curve which may reveal outbursts in brightness or other behaviour. It is quite normal for a comet to be somewhat brighter after perihelion than before, and for the subsequent fade as it retreats to the depths of space to be slower than the rise on the way in.

Estimates are made, as for variable stars (Section 9.1), by comparing the comet with two field stars of known brightness, one brighter and one fainter than the comet. The stars are, of course, tiny point sources whereas the comet is diffuse. This discrepancy can be dealt with by defocusing the telescope or binoculars so that all three appear equally diffuse – the so-called Bobrovnikoff method for making estimates. There are a number of variations of this method, but it is the one most favoured by regular comet observers.

Stars can be identified from *Uranometria 2000.0*, and their magnitudes found in catalogues or computer databases. Many amateur observers use *Sky Catalogue 2000.0* as a standard. Another useful source is the *AAVSO Star Atlas*, designed for variable-star work and giving magnitudes of field stars on the charts themselves, which go down a limit of mag. +9.5. Charts from the *AAVSO Star Atlas* were used as a standard for amateur observers world-wide during the International Halley Watch of 1985–86 (Edberg, 1983).

8.3.3 Drawings: Coma and Tail Measurements

In addition to magnitude estimates, drawings of a comet's visual appearance can be very useful. The experienced French observer Jean-Claude Merlin has made the very valid point that the eye's wider range of contrast perception can allow fine details to be made out in a comet's inner coma; these details are often lost in photographs, which tend to be overexposed for the head region (Merlin, 1994). Carefully made drawings can also be used to derive the coma's diameter and shape, and are an excellent means of recording a comet's tail(s). Drawings are made as "negatives" – bright features drawn dark against a white background (Figure 8.2, *overleaf*).

Many faint comets have little or no visible tail, showing only the coma, perhaps surrounding a point-like "false nucleus"; the true nucleus is never seen, being shrouded by gas and dust emission. It is recommended that observers start their drawing by marking the position of the apparent nucleus relative to the field stars, then fill in the other details around it. The extent and shape of the coma should be sketched in as accurately as possible. A soft pencil and eraser (or an

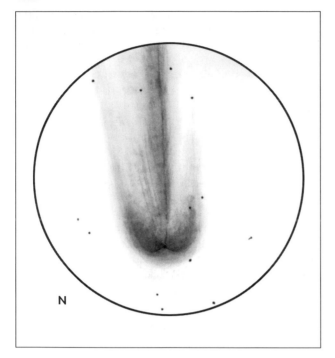

N

Figure 8.2. C/1996 B2 Hyakutake, as seen through 15 × 80 binoculars on 1996 March 23/24, 00:15–01:15 UT. Fans of material can be seen emerging to either side of the near-stellar nucleus, from which a long ion tail extends towards position angle 230°.
Drawing by Rob Bullen

artist's "stub") can be used to render the diffuse outer regions. An observer using, say, a 200-mm telescope at a magnification of ×200 to examine the inner coma may be able to make out structures such as spikes or jets in the brighter, more active comets (as with 109P/Swift–Tuttle in 1992, for example). Careful positioning of these on the sketch may enable subsequent analysts to determine the rotation period of the nucleus.

Tail details may be harder to pick out, especially for fainter comets. The bluish ion tail may have low contrast with the sky background, and its apparent extent will be influenced by the transparency of the atmosphere. Dust tails, where present, are usually easier to see. The observer should try to position the tail(s) accurately relative to the star background on the field drawing. Careful examination of the ion tail can reveal knots or kinks resulting from interaction with the solar wind, while in active comets the dust tail might be striated or banded; such features should be recorded in the drawing. In rare cases, dust ejected in the plane of the comet's orbit can appear as a spiked "antitail" pointing ahead of the comet, as seen most notably for C/1956 R1 Arend–Roland in 1957, and in others such as C/1987 P1

Bradfield in 1987–88. An antitail may be seen when the Earth lies close to the orbital plane of a comet; at other times the dust is seen from the side as a broad sheet or fan.

When making the initial drawing at the telescope or binoculars it is important to work quite quickly, since the comet may move substantially against the background over an interval of 30 minutes or more; as an extreme example, the fast-moving IRAS–Araki–Alcock moved by about 0.6° across the sky in the course of 40 minutes on the nights around its closest approach on 1983 May 10. Most other comets are slower, but the amount by which they move in the 20 minutes or more that it takes to make a drawing may reduce the accuracy of subsequent angular measurements derived from the drawing.

Ideally, the observer makes a rough sketch in the field, with plenty of annotations for later guidance, then – as soon as possible afterwards, so that the details are fresh in the mind – works up a tidy, finished version. Some observers use tracings of the plotted star positions from *Uranometria 2000.0*, in conjunction with their field sketches, to position the features more accurately relative to the stellar background. Indeed, tracing paper is an excellent medium on which to make the finished version of a comet drawing. Such drawings have the advantage of a uniform scale, and can be used to determine the diameter of the coma, and the length and position angle of tails. Position angles are measured relative to the coma from north (000°), through east (090°), and so on. Lines of right ascension on the atlas grid provide a convenient indication of the field's north–south direction. It may be difficult to decide on the position angle to assign to a strongly curved dust tail.

Summary reports of comet observations, together with copies of any field drawings (which will not always be possible, or worth while, particularly for very faint objects at the limits of detection), should be sent to the appropriate section of a national organisation such as the BAA Comet Section or the Comets Section of ALPO. Among the essential details to be given are the time (UT) of the observation, the diameter and magnification of the instrument used for the observation, and an indication of the sky conditions. UT is usually expressed as a decimal of the day; for example, 1998 March 31.86 is the same as March 31, 20:53 UT. Magnitude estimates, coupled with an indication of the degree of condensation of the coma, are very useful. Position angles and lengths (in degrees) for the ion and dust tails, if present, should be given separately.

Reports giving all these details are collected from observing organisations and individuals from around the world by the *International Comet Quarterly*, which compiles and publishes them at three-month intervals. Professional researchers and others regard these compilations as a valuable resource for improving our understanding of the behaviour of comets when close to perihelion.

8.4 Comets in Outburst

One of the most unusual periodic comets is 29P/Schwassmann–Wachmann 1, discovered in 1927 by two German astronomers. The comet follows a 15-year, near-circular orbit somewhat beyond that of Jupiter, and is normally very faint, around mag. +15 to +16. At irregular intervals, and for reasons as yet poorly understood, it undergoes a marked brightening of up to five magnitudes, becoming accessible in, say, a 200-mm reflector. Thanks to its "planetary" orbit the comet can be followed for much of the year, and many observers check its ephemeris position on each possible clear night to see whether an outburst has occurred.

Another comet discovered, in 1930, by the same observers – 73P/Schwassmann–Wachmann 3 – underwent a dramatic brightening on its 1995 apparition. This comet has a more typical elliptical short-period orbit of roughly 5.5 years. In the run-up to its 1995 September perihelion, it had already been noted as brighter than expected from its past behaviour. By mid-October, when it was predicted to be a difficult mag. +13 object, some amateur observers were estimating it to be as bright as mag. +5.0. Detailed images obtained at the European Southern Observatory in Chile revealed the nucleus to have broken into four fragments, accounting for the unexpected brightening. Quite what, if anything, will be seen of the comet around 2006 is anyone's guess.

There are certainly parallels with the famous 3D/Biela, another short-period comet seen to fragment into two nuclei at its return in 1845–46, and as a double comet again in 1852. At the next favourable return, in 1872 November, close passage of the Earth to the comet's expected position resulted in a spectacular meteor shower – the Bielids – produced by debris from the presumably defunct comet. Unfortunately, our orbit does not come close enough to that of 75P/Schwassman–Wachmann 3 for a meteor storm to be produced as a result of the comet's disintegration.

Among the most spectacular comets of the late 20th century was the long-period C/1975 V1 West, which was at its best in February and March of 1976. With a very prominent fan-shaped dust tail, and overall magnitude of –1, West was a fine sight for early-rising observers in the spring pre-dawn. Telescopic observations showed that its nucleus broke into four fragments a few days after its February 25 perihelion, and by the end of March the fragments were quite well separated. Comet West will probably return in instalments, separated by many tens or hundreds of years, in about 100 000 years' time.

Several groups of comets with very similar orbital characteristics have probably had their origins in similar nuclear break-up episodes. Best known are the Kreutz sungrazers, including the bright C/1965 S1 Ikeya–Seki. More recently, C/1996 Q1 Tabur, a good mag. + 5 binocular object seen in the autumn of 1996, was found to be related to C/1988 A1 Liller, which reached mag. + 5.5 in the spring of 1988; perhaps further fragments from the same comet remain to make their appearances in the years ahead.

Perhaps the most surprising outburst of all came from 1P/Halley in 1991. At its most recent return, the comet came to perihelion on 1986 February 9. Observers in the northern hemisphere had a reasonable view of it through the autumn and early winter of 1985–86, when it came close to the fringes of naked-eye visibility and was certainly a good (though not spectacular) binocular object, reaching mag. +5 before heading inaccessibly far south. Close to and immediately following perihelion, 1P/Halley was best seen from the southern hemisphere, showing a fan-shaped dust tail and reaching a peak brightness of mag. +3. The Earth was far from the comet on this occasion, so the return was less impressive than the previous close approach in 1910.

As it retreated from the inner Solar System, 1P/Halley had faded from view as far as amateur observers were concerned by 1987, but professional telescopes continued to follow it outwards. Observations in 1991 February showed the comet to have brightened markedly when at a distance of 14.3 AU from the Sun – well beyond the orbit of Saturn, and well past a distance at which a comet would usually be expected to show activity. The precise nature and cause of the brightening, believed by most to be accounted for by a short-term ejection of dust, remain open to debate. One popular theory is that the comet's surface was somehow disturbed as a high-speed coronal mass ejection (Section 3.1.1) swept by during a particularly vigorous phase of solar activity.

Perhaps other comet nuclei show similar outbursts at great distances from the Sun. Leaving aside the unusually large object 95P/Chiron, only 1P/Halley has ever been followed so far out beyond the inner Solar System.

8.5 Ion-tail Disconnections

One of the most fitting descriptions of comet tails by a professional astronomer must surely be that provided by Fred Whipple, who referred to them as "solar windsocks". Fifty years ago, before spacecraft exploration allowed the Earth's magnetosphere to be mapped out, scientists such as Ludwig Biermann were looking to comets to provide clues as to how the solar wind streaming out from the Sun interacts with bodies in its path. As we have already seen, the solar wind shapes the coma surrounding the nucleus, and draws out the ion tail downwind. Among the more interesting transient phenomena to be witnessed by comet observers are those events where an ion tail which has been pointing in one orientation away from the Sun undergoes *disconnection*, to be replaced by another growing in a slightly different direction.

From studies of disconnection events, Biermann and others constructed models for the solar wind. Building from these models, solar physicists have developed the modern picture in which the solar wind has large-scale sectors of varying magnetic polarity (north or south of the plane of the ecliptic) separated by fairly distinct boundaries. Abrupt changes in the local interplanetary magnetic field (IMF) occur when one of these sector boundaries sweeps past an object. Sector boundary crossings can give added impetus to the Earth's magnetospheric dynamo, sometimes leading to increased auroral activity.

Figure 8.3. C/1996 B2 Hyakutake, showing a disconnection event in its ion tail, photographed from Mt Teide, Tenerife, on 1996 March 25/26 at 00:19–00:23 UT with a Canon 85-mm lens at f/1.8, on hypersensitised Kodak 2415 film.
Courtesy of the observers, Nick James, Martin Mobberley and Glyn Marsh

When a sector boundary sweeps across a comet, the existing ion tail can become pinched off and detached from the coma. A few hours later the disconnected tail may be seen drifting off some way downwind of the comet's head, from which a new ion tail begins to develop at a rather different angle, following the new direction of the IMF. Several such events were seen in 1P/Halley during the 1985–86 apparition, and in the more recent C/1996 B2 Hyakutake of spring 1996 (Figure 8.3).

8.6 The Great Comets of 1996 and 1997

As already discussed, most of the comets that appear in a given year are fairly faint with, at best, only a couple coming within the reach of binoculars. Naked-eye comets are rarer still: observers in the northern hemisphere saw only five (two of which were only marginally naked-eye objects) between 1977 and 1995. IRAS–Araki–Alcock, at mag. +2 during its close (0.031 AU) passage by the Earth in 1983 May, was the brightest. 1P/Halley had just reached naked-eye visibility in 1986 January when it headed southwards and out of view from North America and north-west Europe. C/1987 P1 Bradfield in 1987 November reached mag. +4 while an evening object against the stars of Aquila, while C/1989 X1 Austin was *barely* visible to the naked eye from a dark location as it passed through the stars of Delphinus and Aquila in 1990 May. Rather better was C/1990 K1 Levy, a few months later. By the end of 1990 August it had reached mag. +3, and was quite easy to see in the evening sky as it passed east of Scutum.

Looking back through the historical records, it is easy to see why the second half of the 19th century has been regarded as a golden age for comet observers. During this period several bright comets were seen, including a number which have come to be described as "great comets", thanks to their brilliance and ease of visibility. Among these were the brilliant C/1843 D1, the Great March Comet; C/1858 L1 Donati; C/1860 M1, another "Great Comet"; and C/1861 J1 Tebbutt. C/1877 R1 Coggia and the sungrazing Great September Comet, C/1882 R1, were also spectacular. Each of these could be seen readily by even the most casual skywatcher; by contrast, 1987's Bradfield and Levy in 1990 were noticed as naked-eye objects only by experienced amateur astronomers familiar with the sky.

Throughout the 1980s and early 1990s, many comet enthusiasts bemoaned the lack of great comets in the 20th century, by comparison with the events of a hundred years earlier. To be fair, the 20th century began well, with the brilliant Great January Comet, C/1910 A1, followed by a very favourable apparition of 1P/Halley a few months later. The next three bright comets – C/1927 X1 Skjellerup–Moritsany, the Southern Comet C/1947 X1 and the Eclipse Comet C/1948 V1 – were best seen from southerly latitudes. It was not until the appearance of Arend–Roland, at its best during April and May 1957, and Mrkos in August of the same year, that observers in the northern hemisphere were next treated to a bright comet. The sungrazing Ikeya–Seki, like others of the Kreutz Group, was best seen from southerly latitudes during its brief spectacular apparition in 1965 October. C/1969 Y1 Bennett was a better prospect for those at more northerly latitudes, reaching mag. 0 in 1970 April. The disappointing performance of Kohoutek in early 1974 has been well documented, but West in 1976 did prove to be spectacular.

Arend–Roland, Mrkos, Bennett and West surely each compared favourably with the great comets of the 19th century. Perhaps the apparent dearth of 20th-century great comets is more a matter of semantics, coupled with the long gap between the appearance in northern skies of bright comets between 1910 and 1957, and again for almost twenty years following Comet West (Gingrich, 1995).

8.6.1 Hale–Bopp in 1995

The "drought" was certainly broken with the discovery of C/1995 O1 Hale–Bopp in the summer of 1995. The comet was found independently by two American observers, Alan Hale in Cloudcroft, New Mexico, and Thomas Bopp in Stanfield, Arizona, on the night of 1995 July 22/23. At this time Hale–Bopp was a faint mag. +10 object, appearing close to the mag. +9 globular cluster M70 in Sagittarius. Confirmatory observations soon allowed the comet's orbital path to be more reliably determined, and the prognosis was exciting indeed: Hale–Bopp had been discovered at a distance of 7 AU from the Sun – beyond Jupiter's orbit – and was already unusually bright and active for a comet so far from perihelion. Its high-inclination orbit would bring it to perihelion at 0.9 AU from the Sun, just inside the Earth's orbit, on 1997 April 1 and about 1.3 AU away from us. The prospect of a prominent comet, quite similar in viewing circumstances to Arend–Roland in 1957, encouraged many commentators to propose that this would be a great comet at long last.

Caution was, of course, shown by those who remembered the comparatively poor showing made by Kohoutek in 1973–74. Like Hale–Bopp, Kohoutek had been unusually bright when far from the Sun. In all fairness, Kohoutek did go on to be a respectable binocular object at mag. +4 with a 25° tail. If Hale–Bopp's 1995 July brightness was the result of an unusual outburst, which might die down before perihelion, then it too could prove disappointing.

Imaging from the Hubble Space Telescope during the summer of 1995 showed shells of dust being ejected from the coma of Hale–Bopp, and led to estimates of up to 40 km for the diameter of the nucleus.

At the time of its discovery Hale–Bopp was in the deep south of the sky, at least as far as observers in the British Isles were concerned. Remarkably, David Strange managed to image the comet using a CCD camera attached to a 520-mm reflector from his Worth Hill Observatory in Dorset on the south coast of

England, from where Hale–Bopp at this time could attain an altitude of no more than 7° above the sea horizon. Those in the United States and the southern hemisphere had more favourable views.

Throughout the rest of 1995, Hale–Bopp remained at around mag. +10 and was best covered by observers in the southern hemisphere, for whom it was higher in the sky above the horizon haze. Some reports already suggested the presence of spiral jets of material emerging from the coma. By mid-November the comet was becoming lost in the Sun's glare as it neared conjunction. It remained to be seen how the comet's brightness would increase in 1996, when we would have a better idea of whether this *was* to become a great comet. In the meantime, a welcome diversion completely upstaged Hale–Bopp for a couple of months.

8.6.2 Hyakutake

C/1996 B2 Hyakutake was discovered on 1996 January 30 as a mag. +9.5 object in Libra. The discovery was the second in the space of under six weeks for Yuji Hyakutake, observing from southern Japan using a pair of 25×150 binoculars. The comet was beyond the orbit of Mars at this time, but once its trajectory had been determined it was realised that it would pass close to the Earth – about 0.10 AU distant on the night of March 24/25 – on its way to perihelion. At this time the comet would lie in Draco, between the Plough and Ursa Minor, with the Moon reasonably well out of the way (as a waxing crescent setting in late evening).

The comet's motion across the sky was quite rapid, passing from Libra to Boötes around March 22, and on past Polaris on March 27 and 28. Thereafter, as it receded from the Earth and on towards its May 1 perihelion, it would be in Perseus, as an evening object, for several weeks. When at its brightest, at close approach, Hyakutake would be circumpolar for observers at higher northerly latitudes. The comet was expected to become a prominent naked-eye object around closest approach, then to fade for a time before becoming bright again as it neared perihelion.

Unlike Hale–Bopp, whose coming apparition was anticipated well in advance, Hyakutake's sprint across the northern sky gave observers little time to prepare. An added source of frustration for many in north-west Europe was the poor weather in the spring of 1996, but in the end most observers obtained good views of the comet, which turned out to be the finest since West.

Observers in the British Isles got their first good views of Hyakutake after about March 15. At this time it was still some way out, but had already reached mag. +2.5 and was instantly recognisable as a comet to the naked eye – no need for detailed finder charts in this case! By March 18/19, a tail extending some 3–4° away from the coma to the west (position angle 270°) was evident. A couple of days later the coma could be seen easily with the naked eye, even from quite badly light-polluted localities, as a fuzzy mag. +2 teardrop. There could be little doubt by now that better was to come.

By March 23/24 Hyakutake was becoming truly spectacular (Figure 8.4). Lying not far from Gamma Boötis, the coma was now at least as bright as Arcturus (mag. 0). Even under hazy conditions it was easy to follow the tail for at least 10° with the naked eye, while binoculars showed it to extend much farther. In the early-morning hours the comet stood high in the south-eastern sky, and was a thoroughly memorable sight. Many observers, the author included, took the chance to just sit back and take in the spectacle above from the comfort of a deckchair!

In 10 × 50 binoculars the coma filled about a third of the 5° field of view, with a yellowish oval inner condensation marking the false nucleus. Many observers commented on the bluish and greenish colours in the coma and tail. Still well ahead of

Figure 8.4.
C/1996 B2 Hyakutake high among the stars of Boötes at 00:00 UT on 1996 March 23/24. Captured on Ilford Delta ISO 400 film, using a 20-second exposure with a 50-mm lens at f/1.8; negative print.
Photograph by the author

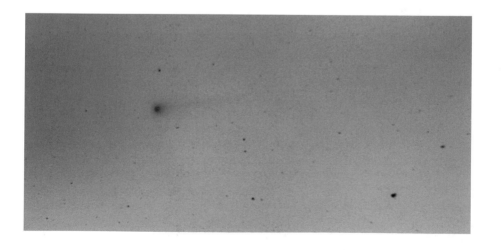

perihelion, the comet did not appear to be emitting large amounts of dust; the magnificent, long, slightly fanned tail seen by observers around Hyakutake's close passage to Earth was its ion tail. This ion tail underwent a major disconnection event on March 24/25. By March 26/27 the coma lay close to Polaris, while the ion tail stretched on through the bowl of the Plough and beyond. Some observers estimated the naked-eye length of the tail from a dark, light-pollution free site, to be more than 40°. The coma remained close to mag. 0.

A few nights later, strong moonlight and the comet's retreat sunwards conspired to reduce the spectacle somewhat, but the ion tail and bow-shaped coma remained impressive in binoculars and small telescopes. The total lunar eclipse of April 3/4 came as something of a bonus for comet observers – a window of about an hour and a half during which the full Moon's glare would be extinguished, allowing Hyakutake, now in the north-west sky against the stars of Perseus, to be observed in detail once more. By mid-totality the comet had sunk quite low in the north for observers in the British Isles. Observers in North America, however, saw the comet rather higher up during the eclipse. By now it had faded somewhat, to about mag. +2.5, and had a shorter fan-shaped tail of perhaps 3–4° length at position angle 070°.

Over the next ten days the comet remained low in the north-west, passing close to Algol on April 12. For many of us the last truly splendid view of Hyakutake came on April 17/18, by which time it had begun to brighten again as perihelion approached. An elongated mag. +2 coma had a long, gently fanning tail of 10° stretching vertically upwards between Algol and Rho Persei as dusk deepened. Dust emission was more in evidence at last, with those parts of the tail close to the coma appearing yellowish. Within a few more days the comet had slipped too far south to be observed from northern latitudes. As it closed on the Sun, it also became lost in the glare. Observers in the southern hemisphere recovered Hyakutake in early May, following perihelion, as a rapidly fading mag. +4 object with a much foreshortened tail.

Indisputably a great comet, Hyakutake whetted observers' appetites for the coming spectacle of Hale–Bopp. Those who started observing comets after 1976 discovered that comets really *could* be spectacular after all.

8.6.3 Hale–Bopp at Its Best

Observers in the southern hemisphere recovered Hale–Bopp after conjunction in mid-February of 1996, by which time it had reached a magnitude of around +9. Over the next few months the comet brightened steadily, reaching mag. +7 by May but remaining a difficult object for observers at more northerly latitudes. A tail – or, more accurately, a dust "fountain", as it was described by veteran American comet observer John Bortle – extended for perhaps 20′ from the coma.

For many in north-west Europe and North America, the summer of 1996 brought the first good views of Hale–Bopp as it tracked gradually westwards and northwards against the stars of Scutum. By July it was passing south of the Wild Duck Cluster (M11), at about mag. +6.5 an easy binocular object. The diffuse coma showed a condensation to its south-east, marking the nuclear region. The brightness continued to rise, to about mag. +5.5 in late August, by which time the comet was 3.5 AU from the Sun, inside the orbit of Jupiter but still farther out than that of Mars.

In 1996 October, as it moved against the star background of Ophiuchus and into the evening sky, Hale–Bopp had reached mag. +5 and was becoming noticeably larger – observers' estimates of the overall diameter were now typically around 20–25′. The tail was beginning to become more prominent, too, fanning out in two sections. To the north-north-east, at position angle 010°, about a degree of tail was visible in binoculars, while a second, shorter tail section extended for about half this distance towards PA 315°.

As 1996 drew to a close, the rate at which Hale–Bopp was brightening appeared to have slowed down. Through the summer the comet had brightened by at least half a magnitude each month, but in the autumn it seemed to have "stalled" at about mag. +5.0, even though it was rapidly closing on the Sun. Was another potential great comet about to let us down? Hale–Bopp became ever more difficult to observe in the evening twilight during November and December as it came to conjunction with the Sun once more. Estimates from mid-December suggested the comet to have reached mag. +4, while the coma and tail continued to grow.

Observers waited in suspense through the first couple of weeks of 1997 for Hale–Bopp to re-emerge as a morning object following conjunction. Through the period of conjunction some were able to spot the comet low in the twilight. Sparse observations from the middle of 1997 January suggested that mag. +3 had been reached, and by the end of the month any fears of another cometary disappointment were well and truly allayed.

For would-be observes in the British Isles, 1997 January was generally a cloudy month, the poor spell of weather finally ending in early February. Hale–Bopp now stood reasonably high in the eastern sky before dawn. Lying just south of the tail of Sagitta's arrow on the morning of February 3, the comet was instantly obvious to the naked eye at mag. +2.5. The tail fanned out towards PA 325° with a length in binoculars of at least 2.5°: the comet was now very much larger than the faint, fuzzy ball that observers had keenly tracked throughout the previous autumn. This was to be the beginning, for many of us, of the comet apparition of a lifetime.

By early March Hale–Bopp had brightened markedly, to mag. +0.8 or brighter; there are few suitable stellar comparisons at this bright end of the range, and most observers' estimates were simply guesses. Early-morning views at the month's opening were breathtaking, revealing a two-component tail, with the narrow ion tail more or less vertical to the horizon and a broad, fanning dust tail spreading out to the west. One excellent description, by a non-astronomer, likened the comet to a shuttlecock hanging in the pre-dawn sky. The tail was now beginning to open out towards us, appearing in its full splendour stretching away from the coma for some 10°. The separation between the ion and dust tails became ever more obvious.

Few comets are amenable to short-exposure photography. Most require long, carefully guided telescopic exposures corrected for the comet's motion against the star background. Hyakutake and Hale–Bopp were exceptions to the rule, and a great many people – astronomers and non-astronomers alike – were able to capture these two objects in short (15–20 seconds at $f/2$ with a standard 50-mm lens) undriven exposures on ISO 400 film, using a tripod-mounted camera. In colour exposures, particularly, Hale–Bopp was remarkably fine, the blue emission from the ion tail contrasting beautifully with the yellow of the dust tail.

During the first week of March Hale–Bopp was strictly an object for early risers, appearing in the north-east against the stars of Cygnus in the last hours before dawn. As the month wore on, the comet gradually began to creep up into the western sky after sunset, becoming obvious even to casual observers. Unlike the previous year, the spring of 1997 was marked by a prolonged spell of good weather for observers in northwest Europe, coinciding with the best period of Hale–Bopp's apparition. Even those who had agonised about the lack of bright comets over the preceding two decades probably became quite blasé about the regular evening presence of the mag. 0 interloper.

Close examination of the dust tail with binoculars showed numerous bands and other structures (Figure 8.5). Some of the most remarkable views of all were obtained by observers who looked closely at the coma through telescopes at moderate powers. The inner coma was dominated by roughly semicircular "shells" of material. The appearance of these features has been explained in terms of the comet's nuclear rotation (believed to have a period of 11.5 hours). When exposed to sunlight, active regions on the nuclear surface emitted jets which carried dusty material into the coma, but almost as soon as the rotation brought these regions round to the night-side of the nucleus, they shut down and emission ceased. As these ejected shells were carried downwind they gave rise to the banding and striation seen in the dust tail.

Figure 8.5. Near-nuclear bands of material visible in the inner coma of C/1995 O1 Hale–Bopp on 1997 April 2 (left) and April 8 (right). These bands are believed to have been thrown off by the nucleus's rotation. CCD images by Nick Quinn

Towards the end of March, when at its brightest, Hale–Bopp was circumpolar for northern observers, with its tail sweeping up towards Cassiopeia. Over the following weeks it continued to dominate the evening sky, tracking gradually across Andromeda and on towards Perseus (Figure 8.6). The dust tail was draped over Algol on the night of April 10/11; the star remained clearly visible, a reminder that comet tails are very tenuous. At its brightest Hale–Bopp reached perhaps mag. –1, with an ion tail stretching for 25° and a prominent dust tail 20° long. The false nucleus was an intensely bright yellow elliptical spot in the depths of the coma. What distinguished Hale–Bopp from Hyakutake and many other comets was its long period of easy visibility, lasting several weeks, thanks to its large distance from the plane of the ecliptic. For many at mid-northern latitudes it remained an evening object until the second week of May, though by then it had faded to about mag. +2. Some observers had kept Hale–Bopp under scrutiny for over 18 months!

As it closed on the Sun in May, Hale–Bopp was lost from view for a time, to be recovered by observers in the southern hemisphere in June. By this time the tail had become foreshortened, and the comet was fading as it began its retreat back into the depths of space. By

Figure 8.6.
C/1995 O1
Hale–Bopp against the stars of Perseus, on 1997 April 7/8 at 20:40 UT; exposure 20 seconds on Ilford HP5plus ISO 400 film, with a 50-mm f/1.8 lens.
Photograph by the author

1998 February, it had faded to mag. +8.5. Recent calculations by Brian Marsden of the IAU Central Bureau for Astronomical Telegrams suggest that Hale–Bopp is on a long-period orbit, the previous return having occurred in 2215 BC; it will return around AD 4389.

Hyakutake and Hale–Bopp certainly brought to an end the drought of great comets in the late 20th century. When the next great comet will appear no one can say. In the interim, however, there will be plenty of lesser objects to observe.

References and Resources

Aguirre, EL, "Comet Hyakutake's spectacular performance". *Sky & Telescope* 92 1 23–30 (1996).

Aguirre, EL, "The great comet of 1997". *Sky & Telescope* 94 1 50–57 (1997).

BAA Comet Section Web Site: http://www.ast.cam.ac.uk/~jds/

Bortle, JE, "Great comets in history". *Sky & Telescope* 93 1 44–50 (1997).

Edberg, SJ, *International Halley Watch Amateur Observers' Manual for Scientific Comet Studies.* Sky Publishing (1983).

Gingrich, M, "Great comets, novae, and Lady Luck". *Sky & Telescope* 89 6 86–9 (1995).

Hendrie, MJ, "The two bright comets of 1957". *Journal of the British Astronomical Association* 106 6 315–30 (1996).

International Comet Quarterly, Smithsonian Astrophysical Observatory, 60 Garden Street, Cambridge, MA 02138, USA.

James, N, "Comet C/1996 B2 (Hyakutake): The great comet of 1996". *Journal of the British Astronomical Association* 108 3 157–71 (1998).

JPL Comet Page: http://encke.jpl.nasa.gov/RecentObs.html

Kronk, G, *Comets: A Descriptive Catalog.* Enslow (1984).

Merlin, JC, "Comets" in *The Observer's Guide to Astronomy*, edited by P Martinez, Vol. 1. Cambridge University Press (1994).

Shanklin, J, "Comets" in *The Observational Amateur Astronomer*, edited by P Moore. Springer (1995).

Chapter 9

Variable Stars

As of 1998, at least 30 000 stars have been identified and catalogued as varying in brightness, while about half as many again are suspected of variability. In most instances the variations are either too slow (of the order of months, years or decades) or too subtle (fractions of a magnitude) to be included in any discussion of readily observable transient astronomical phenomena. None the less, there is much value in systematic, long-term series of observations of slowly varying red semiregular variables or long-period (Mira) stars by amateurs.

Sometimes even ostensibly well-studied stars can throw up surprises. Mira (Omicron Ceti) itself, for example, comes to maximum light fairly predictably at intervals of about 330 days, and is a favourite target for many amateur observers. At maximum it is usually a reasonably easy mag. +3 naked-eye object to be found a few degrees west of the triangle of stars marked out by Alpha, Gamma and Delta Ceti. There can, however, be considerable variation in the peak brightness from one maximum to the next: sometimes Mira struggles to be brighter than mag. +5, while at other times – as in early 1997 – it can be surprisingly bright, around mag. +2. Long-term trends in the behaviour of such stars are revealed only if observations spanning several decades are available for analysis. The archives of the BAA Variable Star Section (BAAVSS) and American Association of Variable Star Observers (AAVSO), which stretch back many years, come very much into their own in such studies.

While slowly changing red variables make up the bulk of some amateur observing programmes, there are a number of other types whose variations are more rapid, in particular the eclipsing binaries and eruptive variables.

9.1 Eclipsing Binaries

Many of the apparently lone stars in the night sky are members of close double stars. Some can be resolved with small telescopes, and double star observing is a much enjoyed aspect of deep-sky observing. When the two components of a binary system are too close to be resolved even in powerful instruments, their binary nature may be revealed by subtle periodic changes in their combined spectrum. In systems where the orbital plane is suitably oriented with respect to the Solar System, stars orbiting around a common centre of gravity will pass in front of one another at regular intervals. Where the members of a pair are of unequal magnitude, their alignment leads to periodic eclipses, during which the combined light of the system drops. These systems are known as *eclipsing binaries*.

The best-known eclipsing binary is Algol (Beta Persei), consisting of a bright spectral-class B8 star about a hundred times as luminous as the Sun, and a fainter K2-class subgiant, separated by 10.4 million km and lying at a distance of 92 light years from the Solar System. At intervals of about 2.87 days the dimmer component passes in front of the brighter across our line of sight, causing a fade from mag. +2.1 to +3.4, quite noticeable to the naked eye.

Careful examination of Algol's light variations using sensitive photoelectric equipment shows that the system actually has two minima in each cycle. Between the deep, readily observable primary minima there are secondary minima, not detectable by eye, in which the combined magnitude of the system dips by 0.05 mag. as the brighter star passes in front of the fainter.

Tracking Algol's primary eclipses is a very worthwhile exercise, and one which a surprising number of even quite experienced amateur observers have never carried out. The observing schedule is quite straightforward, and once it is completed the observer will have a satisfying light curve to show.

While apparently quite predictable, Algol has been found, through careful analysis of observations collected by the Junior Astronomical Society in the 1970s and 1980s, to be slowing down (Isles, 1991a). Minima of the system were found to be occurring late by about a second, with a suggested precise period from minimum to minimum of 2.867315 days, as opposed to the ephemeris value of 2.8679 days. The deviations are believed to stem from either mass exchange between the two stars, or perhaps the ejection of material from the system. These findings underline the value of even naked-eye observations, provided that they are carefully made. Discoveries can be made using the very simplest equipment.

Most of the eclipsing binaries for which timings of minima can be usefully made undergo their fade over the course of 6–12 hours. For observers at northerly latitudes, Algol can be seen to undergo its entire 10-hour primary eclipse cycle in the course of a single night around midwinter. Forecasts of the times of primary eclipse are given in annual handbooks, and can be used as a rough guide as to when minimum will occur. The popular magazines also highlight particularly favourable eclipses.

Ideally, magnitude estimates should be made at intervals of 15–20 minutes for a period of at least 4 hours either side of the expected minimum. Visual magnitude estimates are made using non-variable comparison stars, some brighter and some fainter than the variable. Many observers find the *fractional method* (Isles, 1990) most convenient for estimating magnitudes. In this method the variable's magnitude is derived by simple arithmetic from its difference from those of the comparison stars. Suitable comparisons for Algol are listed in Table 9.1 (*overleaf*).

An example of the use of the fractional method is provided by an eclipse of Algol which the author followed on Christmas night, 1976. The magnitude at 18:50 UT was estimated using Alpha Cephei (mag. +2.43) and Epsilon Cassiopeiae (mag. +3.38) as comparisons. Dividing the apparent difference between the comparisons into five steps (a number most observers find comfortable for estimation purposes), Algol was estimated to lie three steps fainter than Alpha Cephei and two steps brighter than Epsilon Cassiopeiae. This was logged as

Alpha Cep (3) V (2) Epsilon Cas

Table 9.1. Comparison stars for Algol

Comparison Star	Magnitude
Alpha Persei	1.80
Alpha Andromedae	2.07
Beta Cassiopeiae	2.26
Alpha Cephei	2.43
Alpha Pegasi	2.48
Gamma Pegasi	2.83
Epsilon Persei	2.89
Delta Persei	3.03
Eta Aurigae	3.17
Epsilon Cassiopeiae	3.38
Alpha Trianguli	3.44
41 Arietis	3.63
Kappa Persei	3.81

It is then a matter of straightforward arithmetic to arrive at the magnitude for the variable (V), Algol. The magnitude difference between the comparisons is 3.38 – 2.43 = 0.95 mag., so each step is equal to 0.95/5 = 0.19. Algol's magnitude is 3 × 0.19 = 0.57 mag. fainter than Alpha Cephei, i.e. mag. 2.43 + 0.57 = mag. +3.00.

It should be noted that, while an accuracy of two decimal places is used in the arithmetic, the eye is incapable of detecting magnitude differences of more than 0.1 or 0.2 mag. at best, and variable-star estimates should therefore always be rounded to the nearest 0.1 magnitude. The above estimate should therefore be stated as mag. +3.0.

Plotted as a light curve of magnitude against time, estimates made on a given night can be used to gauge the time of minimum. Figure 9.1, based on the author's observing log, shows the eclipse of Algol on 1976 December 25/26. According to the 1976 BAA *Handbook*, minimum on this night was expected at 20.3^h UT, or 8.3^h Greenwich Mean Astronomical Time (GMAT, which is equivalent to UT minus 12 hours, is used as a standard in variable-star work largely to avoid ambiguities arising from the change of date at midnight during night-long observing runs). Observations were commenced at 17:55 UT ($5^h 55^m$ GMAT); the resulting light curve yields an estimate (to the nearest 10 minutes, a reasonable degree of accuracy) for the time of minimum of around 20:10 UT ($8^h 10^m$ GMAT) – in good agreement with the ephemeris.

A better light curve can be assembled if several independent observers make estimates on the same

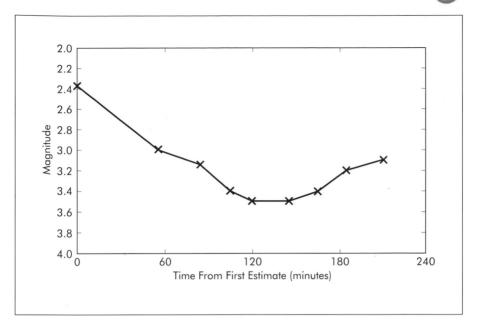

Figure 9.1. Light curve for the Algol eclipse of 1976 December 25/26, based on estimates from the author's observing log.

night, and pool their data. On most occasions the duration of the whole eclipse does not coincide with the hours of darkness, and only the ascending or descending stage can be covered. Observations on these occasions, combined with those from other nights, can still be used to define the time of minimum relative to the ephemeris value. The estimates, pooled with those of other observers, are plotted as a function of phase (from 0, the predicted time of minimum, to 1, the time of next minimum) and can be built up into a cumulative curve. Obviously, only observations made within a few weeks of one another should be combined in this way. Such plots can show whether or not stars are performing according to their ephemerides, and offer greater accuracy in assessing long-term behaviour than is possible with a single night's observations like those shown in Figure 9.1.

Algol is certainly the best-known eclipsing binary, and probably the most observed, but there are many other examples. Not far from Algol in the sky is another naked-eye eclipsing binary, Lambda Tauri, some 6 or 7° west of the distinctive "V" of the Hyades. Lambda Tauri varies between mag. +3.5 and +4.0, with minima at intervals of 3.953 days. This system consists of a bright B-class primary with a fainter companion 13.6 million km distant. Like Algol, the star was found by JAS observations to be running slow relative to the ephemeris, by about 3 seconds (Isles, 1991b). Predictions of minima appear in the annual BAA *Handbook*, and in *Sky & Telescope*.

Lambda Tauri and Algol are the brightest short-period eclipsing binaries. There are many others whose variations fall within binocular range. The BAA Variable Star Section has an observing programme covering dozens of such stars, and issues predictions via its regular newsletter. Many are stars which have not been observed for many years, and are now found to deviate markedly from their listed ephemerides. There is a great deal of valuable work to be done in helping to update the elements for such stars by observing and timing their minima at the current epoch. Changes can come about as a result of mass exchange, as suspected for Algol, or as a result of precession of the system's orbit, resulting from the gravitational influence of a third, unseen star.

One of the more popular binocular eclipsing binaries is RZ Cassiopeiae (RA 02^h 48.9^m, dec. +69° 38.1′), which varies between mag. +6.4 and +7.8 in a period of 1.195 days. This variable is easily located a few degrees from Iota Cassiopeiae, just beyond the eastern end of the distinctive "W" of the constellation (Figure 9.2). Primary eclipses show a rapid, 2-hour fade and an equally fast recovery (Figure 9.3). Favourable minima can therefore easily be observed in the course of a single evening.

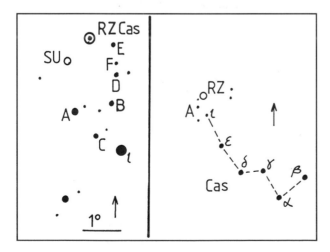

Figure 9.2. The RZ Cassiopeiae field. Comparison star magnitudes: A +6.0; B +6.8; C +7.3; D +7.4; E +7.7; F +8.0. *Drawn by Melvyn Taylor*

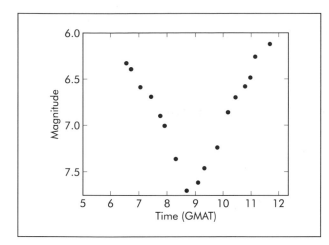

Figure 9.3. Light curve for a minimum of RZ Cassiopeiae, observed on 1997 December 2/3. From it the time of minimum can be estimated to be about 8^h 55^m GMAT. *Courtesy of Melvyn Taylor*

9.2 Cataclysmic Variables

Cataclysmic variables, of which U Geminorum is regarded as the prototype, are extremely close binary systems (with orbital periods of only a few hours) in which one partner is more evolved than the other. The less evolved partner has an extended atmosphere from which material is drawn into an accretion disk surrounding the other component, a white dwarf. At intervals of several tens of days, the mass of accreted material becomes sufficient to initiate fusion reactions. Ignition of this "nuclear runaway" causes the system to flare in brightness by perhaps five magnitudes. The outburst may last for 10–20 days before the system subsides to quiescence; continuing accretion leads to successive rounds of activity. About 250 such stars are currently known.

A similar mechanism of mass exchange in close binary systems, leading to eruptive activity, is understood to operate in novae (Chapter 10); indeed, cataclysmic variables are often described as *dwarf novae*.

The brightest example is SS Cygni (RA 21^h 42.7^m, dec. +45° 35.1′), found near 75 Cygni, some 15° east of Deneb (Figure 9.4, *overleaf*). The components of SS Cygni appear to be a yellow dwarf and more compact white dwarf orbiting around their common centre of gravity in about 6.5 hours. SS Cygni has been monitored continually by AAVSO and BAA Variable Star Section observers since 1896. As shown in Figure 9.5 (*overleaf*), outbursts may be either long (about 18 days

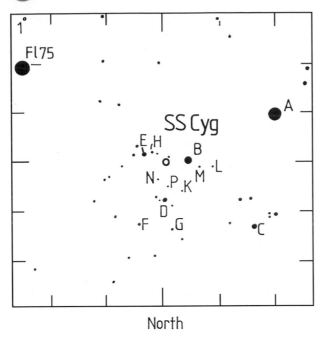

North

Figure 9.4. 1° field chart (inverted) for SS Cygni. Comparison star magnitudes: A +7.80; B +8.72; C +9.16; D +9.78; E +9.92; F +10.36; G +10.89; H +11.00; K +11.19; L +11.50; M +11.54; N +12.00; P +12.41. *Drawn by Melvyn Taylor*

in duration) or short (8 days); long and short outbursts often alternate. Sometimes there are anomalous long outbursts, marked by a slower rise to maximum than the normal 1–2 days. On average, outbursts take place about every 50 days, but the timing of onset is unpredictable. In its quiescent phase, SS Cygni is of mag. +12.4, typically reaching mag. +7.7 (well within binocular range) at outburst. To follow SS Cygni throughout its magnitude range requires a telescope of 200 mm aperture.

Figure 9.5. Light curve for the cataclysmic variable SS Cygni during 1984, derived from BAA Variable Star Section data. *By kind permission of the Director of the BAAVSS, Gary Poyner*

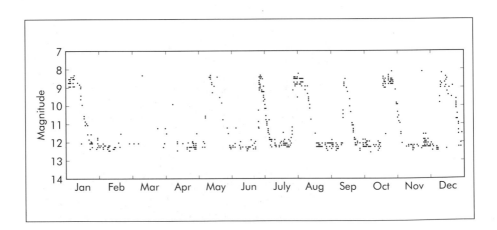

The type star of the class, U Geminorum (RA 07^h 55.1^m, dec. +45° 35.1′), is fainter, with peak brightness in outburst of mag. +8.2, and a minimum of mag. +14.9. The average interval between outbursts of U Geminorum is longer (about 105 days) than for SS Cygni. Short outbursts last about 10 days, long outbursts 16 days.

Like SS Cygni, U Geminorum is a very close pair with the remarkably short orbital period of 4.5 hours. The system can be observed in quiescence to be an eclipsing binary, and is believed to comprise a G-type dwarf and a white dwarf.

9.3 Recurrent Novae

Intermediate between the cataclysmic variables, with their outbursts recurring on timescales of weeks, and the true novae, with outbursts separated by thousands of years, are those variable stars classified as recurrent novae. Again, these are close binary systems in which mass exchange occurs. As with the cataclysmic variables, the eruptive mechanism is believed to be the continued accretion of material from the extended atmosphere of the larger partner onto the dwarf member of the system, leading to nuclear runaway and rapid, short-term brightening.

Perhaps the best-known object in this category is the Blaze Star, T Coronae Borealis, seen in outburst in 1866 and 1946 as an additional naked-eye star near Epsilon Coronae Borealis, just to the east of the circlet of stars marking the "crown" of the constellation. The outburst of 1866 May 12 took the star to a peak brightness of mag. +2.0, from where it faded to below naked-eye visibility in a matter of 10 days. Within a month it had faded back to its normal minimum of about mag. +10.8. Around 100 days after the initial, extremely rapid outburst a further secondary maximum, lasting for about 3 months, took it back to mag. +8.0 before it again faded to minimum.

T Coronae Borealis remained faint for the next 80 years, then was again seen in outburst at mag. +3.0 on 1946 February 9. As in the previous outburst, the peak was short-lived and the star faded rapidly from naked-eye view. The 100-day "bounce" to mag. +8.0 was also repeated.

The components of the T Coronae Borealis system are a red giant and a blue dwarf star. The dwarf's gravitational influence pulls material from, and distorts, the atmosphere of the red giant. The orbital period of the system is 227.6 days.

Only the two outbursts mentioned above have been recorded, so any postulated 80-year periodicity (in which case the next outburst can be expected in 2026) should be treated with caution – another outburst could occur at any time. Many variable-star observers check the T Coronae Borealis field on every possible clear night. In 1998 March it was still around minimum light, at mag. +10.2.

Several other recurrent novae are known which, while fainter at maximum than T Coronae Borealis, appear to undergo more frequent outbursts. Among the best-followed is RS Ophiuchi, which lies a few degrees north-west of the third-magnitude star Nu Ophiuchi, to the west of Scutum. RS Ophiuchi has been seen in outburst five times since 1898. In 1933 August it reached mag. +4.3, while in 1958 July it peaked at mag. +5.0. During the rapid fade in this outburst it appeared markedly red as a result of hydrogen-alpha emission, while spectroscopic studies revealed an expanding shell of material surrounding the system. Most recently, in 1985 January RS Ophiuchi attained mag. +5.3 during outburst, fading rapidly thereafter. Like T Coronae Borealis, RS Ophiuchi merits nightly checking. In outburst it will certainly be visible in binoculars, if not faintly to the naked eye. At minimum it shows some degree of variability around mag. +10 to +11, due to pulsation of the red giant component.

Several other recurrent novae whose maxima take them within binocular range are listed in Table 9.2.

9.4 R Coronae Borealis Stars

Classed as eruptive stars, these are objects which are prone to sudden, unpredictable fades in brightness of up to 8 magnitudes over the course of a few weeks. Recovery to the more normal maximum brightness may take several months, or even years in extreme cases. Several of these stars may go for many years without showing a fade.

The prototype is a favourite star with many amateur observers, and is one of the two brightest of the class. *R Coronae Borealis* shows minor variations around mag.

Table 9.2. Recurrent novae

Star	RA	Dec.	Magnitude Range	Observed Outbursts
T Pyx	09^h 41.7^m	32° 22.7′	6.5–15.3	1890, 1902, 1920, 1944, 1967
T CrB	15^h 59.5^m	+25° 55.1′	2.0–10.8	1866, 1946
U Sco	16^h 22.0^m	−17° 53′	8.7–19.3	1863, 1906, 1936, 1979
RS Oph	17^h 50.2^m	−06° 42.6′	4.3–12.5	1898, 1923, 1958, 1967, 1985
V1017 Sgr	18^h 32.1^m	−29° 22′	6.2–14	1901, 1919, 1973
WZ Sge	20^h 07.6^m	+17° 42.2′	7.0–15.5	1913, 1946, 1978
VY Aqr	21^h 12.2^m	−08° 49.6′	8.0–16.6	1907, 1962, 1973, 1987

+6.0 when at maximum light and is usually easily found in binoculars, within the "crown" of Corona, towards its eastern side (Figure 9.6, *overleaf*). The star is prone to rapid fades: in the early phase of a major dimming it may appear initially about 0.5 mag. fainter than normal – a drop which is quite obvious to those who have been following it closely – before taking a sharp plunge from binocular view over the course of a week or so. The star has earned the nickname of the 'Reverse Nova': rather than seeing a new star appear where none was previously visible, we see a familiar object disappear.

Deep minima can take R Coronae Borealis to as faint as mag. +14, visible only in telescopes of at least 300 mm aperture. Minimum can last for as long as 8 months, as in 1977, before the star recovers to its normal maximum brightness. Most often, recovery to mag. +6.0 takes 4 months or so. Sometimes there are "stalls" in the recovery, showing as secondary minima. A very unusual period of activity saw the star undergo a series of minima, without returning to its usual full brightness, between mid-1963 and early 1967.

Equally, there have been long spells, as between 1925 and 1935, when R Coronae Borealis has remained at maximum. Not all fades are extremely deep: minor dips to no fainter than mag. +8.0 can also occur, as in the autumn of 1989 (Figure 9.7, *overleaf*).

R Coronae Borealis stars are old, high-luminosity objects which are depleted in hydrogen but rich in later

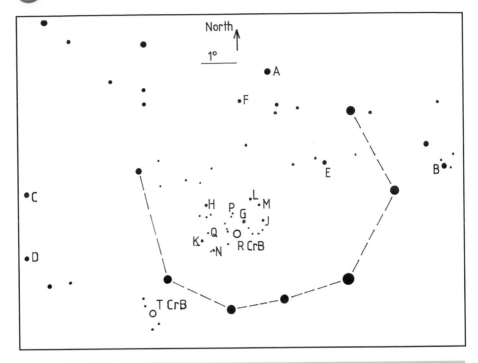

Figure 9.6. Binocular field for R Coronae Borealis. Comparison star magnitudes:
A +5.56; B +5.63; C +5.94; D +6.28; E +6.57; F +6.72; G +7.18; H +7.63; J +7.93;
K +8.28; L +8.60; M +8.88; N +9.20; P +9.36; Q +9.49. The position of the recurrent
nova T Coronae Borealis is also shown.
Drawn by Melvyn Taylor

Figure 9.7. Light curve for R Coronae Borealis between 1987 and 1991, derived from
BAA Variable Star Section data. For much of the time the star is at maximum light (around
mag. +6), but it can undergo dramatic fades to mag. +12, as in the second half of 1988.
In 1989 a less severe, shorter-duration fade occurred.
By kind permission of the BAAVSS Director, Gary Poyner

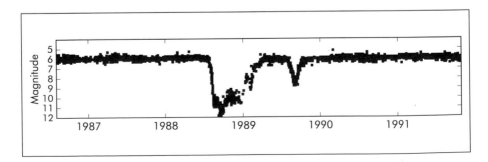

products of nuclear burning – principally helium and carbon. They undergo low-amplitude fluctuations, visible at maximum. The observed fades result from the ejection of clouds of carbon into their atmospheres, followed by its condensation – probably as grains of graphite – which can cause obscuration in our line of sight. According to results from the Infrared Astronomy Satellite, R Coronae Borealis has surrounded itself with a tenuous cloud of carbon extending for 25 light years.

About 30 R Coronae Borealis stars are known, of which the prototype is the most intensely followed. Another well-observed example which is particularly favourable for those at more southerly latitudes is RY Sagittarii (RA 19^h 16.5^m, dec. $-33° 31'$), which varies between mag. +5.8 and +14.0. Many of the others are much fainter at maximum, and impossible to follow through the whole of their light variations.

References and Resources

American Association of Variable Star Observers (AAVSO), 25 Birch Street, Cambridge, Massachusetts 02138, USA.

AAVSO Web Site: http://www.aavso.org

BAA Variable Star Section Web Site: http://www.telf-ast.demon.co.uk

Isles, JE, *Variable Stars*, Webb Society Handbook Vol. 8. Enslow (1990).

Isles, J, "A nova patrol for the JAS?" *JAS News Circular 163* (1991a).

Isles, JE, "Eclipsing binaries, Pegasus to Sagittarius in 1972–1987". *Journal of the British Astronomical Association* 101 4 219–22 (1991b).

Isles, J, "The dwarf nova U Geminorum". *Sky & Telescope* 94 6 98–9 (1997).

Levy, D, *Observing Variable Stars: A Guide for the Beginner.* Cambridge University Press (1989).

Taylor, MD, "Variable Stars" in *The Observational Amateur Astronomer*, edited by P Moore. Springer (1995).

Chapter 10

Novae and Supernovae

The eruptions of variable stars such as SS Cygni and T Coronae Borealis can certainly bring about an impressive increase in their apparent brightness. Some of these objects may undergo brightenings by a factor of more than a thousand. More impressive still are the enormous increases in brightness of novae and, particularly, supernovae.

10.1 Novae

The weekend of 1975 August 29–30 lives long in the memories of those amateur astronomers fortunate enough to have had clear skies. As dusk fell on Saturday August 29, observers heading out to take advantage of a dark evening, ahead of the early rising last quarter Moon, noticed a bright addition to the familiar pattern of Cygnus, high overhead from northerly latitudes. This being a time of year when, particularly at such latitudes, artificial satellites are commonly seen, many assumed that the second-magnitude interloper, about 5° north of Deneb, was just another piece of orbiting "junk". Looking back a few minutes later to find that the object had not moved, observers quickly cottoned on to the fact that this was no satellite. Rather, they were witnessing the eruption of the brightest nova seen for over thirty years.

A quick telephone call to one of the astronomical clearing-houses of the day (these were pre-Internet times!) soon dispelled any thoughts of the glory of discovery. Lines to *The Astronomer* magazine, for example, were jammed as scores of independent observers called to report their sightings. Credit for discovering Nova Cygni 1975 actually went to the Japanese observer Kentaro Osada, who made the first report of the outburst around 12^h UT on August 29, at which time the star was at mag. +3.0; pre-discovery photographs from as little as 12 hours earlier show the nova on the rise, but still fainter than magnitude +10. The nova's rise was extremely rapid: by nightfall over western Europe, some 10 hours later, it had reached mag. +2.

The nova peaked in brightness a little above Deneb, at mag. +1.2 on August 30. Within a matter of days a rapid decline had set in, and by mid-September Nova Cygni was a binocular object. Now designated V1500 Cygni, this was a fast nova, peaking in brightness within a couple of days of discovery, then fading from naked-eye view extremely rapidly (Figure 10.1). Anyone unlucky enough to have had a cloudy week in late August to early September would have missed it completely.

The majority of novae are comparatively faint, perhaps becoming binocular objects at best. Nova Cygni 1975 is one of only a handful of novae to have become an easy naked-eye object in the 20th century. Others are Nova Persei 1901 (mag. +0.2), Nova Aquilae 1918 (peak magnitude −1.4, the brightest for over 300 years), Nova Cygni 1920 (mag. +1.8), Nova Herculis 1934 (mag. +1.3) and Nova Puppis 1942 (mag. +0.3).

Figure 10.1. Light curve for the very fast Nova Cygni 1975.

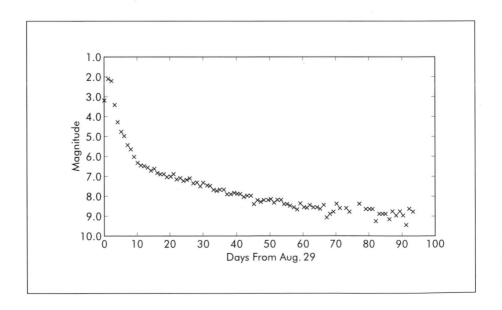

Nova eruptions are believed to occur in close binary systems in which material is exchanged between the components (usually a white dwarf and a red giant). The eruptions may recur over timescales of thousands or tens of thousands of years, and typical novae may be seen as extreme examples of cataclysmic variable stars (Section 9.2). The accretion of material onto the surface of the white dwarf member of the system leads to an extreme nuclear runaway and enormous brightening. Material is ejected at high velocities, as has been determined from spectroscopic observations. Indeed, it is in providing the initial alert of a nova eruption, allowing professional astronomers to obtain spectra during the early stages of the outburst, that amateur astronomers can make their most valuable contribution in this field.

In rare cases, some time following their eruption, novae may be seen, in high-powered instruments, to be surrounded by a faint shell of ejected material. Noted examples are Nova Persei 1901 (catalogued as the variable star GK Per; both the star and the surrounding nebulosity continue to show variations) and Nova Cygni of 1992, the latter's expansion having been imaged by the Hubble Space Telescope.

Nova Cygni 1975 was exceptional in becoming a prominent naked-eye object. It also seems to have been unusual in that deep exposures of the field show no obvious progenitor object, down to the very faint limit of mag. +21, suggesting an extremely large magnitude range; most novae brighten by about nine magnitudes and have progenitor stars which can be identified on images taken many years previously. A number of theories have been put forward to account for the extreme range of Nova Cygni 1975. These include the suggestion that the star was undergoing its first ever, and thus most severe, outburst, based on the probability that novae are recurrent over very long timescales. Another alternative, considered rather unlikely, is that the nova resulted from a solitary, massive star accreting interstellar material, leading to nuclear runaway.

10.1.1 Observing Novae

Following the discovery of a nova, the most useful work the amateur can do is to make magnitude estimates, allowing a light curve to assembled. Estimates are made in exactly the same way as for variable stars (Section 9.1). Where sufficiently early warning has been obtained of a nova on the rise, it may be worth while making estimates on an hourly basis. Observers are certainly encouraged, when possible, to make estimates at least a couple of times each night – perhaps once in the evening, then again during the morning hours – for novae in outburst.

Obviously, in order to make such observations the observer has to know that a nova *is* in outburst. First notification is usually by an IAU *Circular*, which will provide an accurate position for the object. Amateur astronomers will normally be alerted via circulars issued by the major national amateur astronomical organisations, which may also provide finder/comparison charts. Paper circulars are obviously prone to some element of delay, however, and the observer may miss out on making estimates during the critical first few days after the outburst if reliant solely on this mode of communication. In recent years, electronic circulars issued by such organisations as *The Astronomer* magazine have proved to be a very efficient means of getting nova and other alerts to those observers who can best use them. As the Internet has grown in sophistication, it has also become possible to transmit encoded charts, allowing estimates to be made straight away. In the absence of specially prepared charts the observer can make use of those in, for example, the *AAVSO Variable Star Atlas*.

The light curve derived from such observations can be informative. Nova Cygni 1975 can be cited as a perfect example of a fast nova; others show longer, slower outbursts. Probably the best-documented recent instance of a slow nova was George Alcock's Nova Delphini (now designated HR Delphini) of 1967 July, which rose quite gradually, over about a month, from its discovery magnitude of +6.7 to an initial peak around mag. +5. The nova stayed close to this magnitude, with minor dips, for about four months, before rising again to mag. +3.5 – an easy naked-eye object in late 1967 December. Following this peak, a final, slow fade set in, but Nova Delphini remained an easy binocular star for almost three years. In 1997, some 30 years from outburst, HR Delphini is still visible in reasonably large amateur telescopes as a mag. +12 object; pre-discovery photographs of the field revealed the progenitor star to be of this magnitude, and HR Delphini has presumably now returned to its quiescent state.

The light curves of some novae, as they fade, show a marked dip, following which the star brightens again, to continue fading at the original rate. This is sometimes referred to as the DQ Herculis profile, having first been recorded in detail for J.P.M. Prentice's Nova Herculis of 1934. The cause of the dip appears to be the condensation of dusty material in the expanding shell around the nova, causing a temporary obscuration similar to that observed in R Coronae Borealis stars (Section 9.3). Eventually, the dust is cleared by radiation from the nova, leading to the apparent secondary brightening.

10.1.2 Visual Nova Patrols

Many nova discoveries are serendipitous. Nova Herculis in 1934, for example, was discovered during a watch for the Geminids by the distinguished meteor observer J.P.M. Prentice. More systematic searches have been carried out with considerable success by dedicated observers. As with comet-hunting, the visual search for novae demands many hundreds of hours of familiarisation time, during which the observer will sweep the sky with binoculars or a low-power, rich-field telescope. Regular observation of the starfields leads to an intimate, almost instinctive knowledge of their patterns, and experienced observers can recognise any new object instantly as a disruption to a familiar pattern.

The most renowned 20th-century visual nova searcher must surely be George Alcock, who has carried out his work from the village of Farcet, near Peterborough, in East Anglia. Alcock began his observing career in meteor triangulation, working with Prentice in the 1930s, but abandoned this work to begin his sky patrols in the early 1950s. Using a pair of 25×150 binoculars, Alcock has memorised the positions of 30 000 stars and 500 nebulae, and in the course of his searches has discovered five comets (most recently IRAS–Araki–Alcock in 1983) and five novae. Among his nova discoveries were the slow Nova Delphini on 1967 July 8, and Nova Herculis on 1991 March 25. Another observer who has very effectively used binoculars for visual nova patrolling is the Californian amateur Peter Collins, whose discoveries include Nova Cygni 1992.

Committing the whole of the visible binocular sky to memory is an enormous undertaking. Few, indeed, are the observers with the necessary level of dedication to do so, and with the advent of more convenient photographic media it seems likely that visual nova searches will in future become more limited. For those wishing to take up the challenge, the most promising approach now seems to be to restrict the patrol to a limited number of starfields. Visual nova patrollers collaborate in the work of the UK Nova/Supernova Patrol (many of whose members are based outside the UK) operated jointly by *The Astronomer* magazine (TA) and the BAA. Individual observers are allocated a region of the sky, which should be surveyed using binoculars or a rich-field telescope on every possible clear night. A pair of 10×50 binoculars, for example, will allow stars to mag. +9 to be detected, while the larger – and heavier, so necessarily tripod-mounted – 15×80 format will allow objects down to mag. +10.5, or fainter, to be seen.

Since observable nova eruptions occur in binary systems within our own Galaxy, it is hardly surprising that most appear close to the Milky Way. The rich starfields of Cygnus and Vulpecula, and those near the heart of the Galaxy in Scorpius and Sagittarius, are regarded as a particularly happy hunting ground. The main difficulty, in these star-packed parts of the sky, lies in picking out a possible newcomer against the background.

Visual nova-hunting will certainly test the observer's patience, and is not without its pitfalls. Many suspected novae will turn out, on closer inspection, to be long-period (Mira-type) variable stars at maximum, or asteroids traversing the field. As important as dedication and persistence is access to a good star atlas and catalogue providing positional and other details for such variables, and ephemerides for asteroids, particularly those which can become brighter than 10th magnitude. Ephemerides for brighter asteroids are given in annual publications such as the BAA *Handbook* or the RAS of Canada *Observer's Handbook*. Computer-based ephemerides such as those available with *Asteroid Pro* allow rapid checking and the production of field charts. Positions for long-period variable stars which attain a peak brightness of mag. +9 or brighter are shown in atlases such as *Uranometria 2000.0*.

10.1.3 Photographic Nova Patrols

Photographic nova patrols make similar demands on the patience and dedication of the observer, but have the advantage that permanent records are generated from which there is the chance of extracting unexpected additional information at a later date. For example, Mike Collins, one of the BAA/TA team's leading contributors, has discovered more than 130 previously unknown variable stars on patrol photographs taken since the mid-1980s.

Modern, fast black-and-white emulsions are ideal for this sort of work. Kodak T-Max 400 is suitable for push-processing to speeds up to ISO 1000 without substantial degradation of the image by grain formation, allowing exposures of around 60 seconds to be made, which is sufficient to capture stars to around magnitude +10. Many novae are discovered photographically when near this magnitude. Such exposures can be taken readily with a camera piggybacked on an equatorially mounted telescope, or on its own small free-standing equatorial drive. Guiding accuracy for such short, wide-field exposures is not so critical as for the deep imaging of, say, distant galaxies.

The choice of camera for patrol work is a critical influence on success. The observer might instinctively opt for a comparatively wide field of view, so that a single exposure covers a large area of the Milky Way. A standard 50-mm lens, for instance, will give a field of $40° \times 27°$ on a 35-mm format negative. There is, however, a trade-off: on even a short exposure the wide field will be crowded with stars, and it will be difficult to pick out any newcomer. A more fruitful approach is to use a lens of longer focal length; the smaller field of view is more than compensated for by the larger image scale. Observers working under the umbrella of the TA/BAA patrol are each allocated a number of fields measuring $15° \times 10°$, equivalent to the coverage given by a 135-mm lens on 35-mm film. A lens of 200 mm focal length gives an even smaller field ($10° \times 7°$) and larger image scale, but twice as many exposures will need to be made to get the coverage of a 135-mm lens.

For each patrol field, duplicate exposures should be taken in quick succession to avoid possible problems with spurious suspects resulting from flaws in the emulsion. To save time – and allow more exposures to be made on a given night – two identical cameras may be used on the same mount simultaneously; this method has been adopted by, for example, Mike Collins.

Exposures should be inspected as soon as possible after they are taken. For many observers this means processing the film within minutes of taking the exposure. Care should be taken, amid any haste, to ensure that the negatives are properly fixed and dried before inspection, avoiding scratches or tears in the emulsion, and also allowing their long-term preservation for later re-inspection. Individual frames can be stored in slide mounts. A careful record should be kept of each exposure; properly filed and catalogued, an observer's collection of nova patrol exposures may prove a valuable resource for future studies.

Before the observer can really start on photographic nova searching, a "reference" frame of each patrol field needs to be taken against which subsequent exposures are checked. The usual method for comparing reference and patrol exposures is to mount them side by side on an improvised (or commercially obtained) light table, usually featuring an opaque Perspex panel diffusely illuminated from below, so that the two can be observed simultaneously. Standard 35-mm negatives can be examined using a low-magnification hand-held viewer, available from photographic suppliers. Many ingenious blink comparators, in which alternate viewing of the frames will cause a new object to appear and disappear, have been devised by amateur observers. One system, Problicom, which employs two slide projectors, has been used to great effect by William Liller at Villa del Mar in Chile.

Liller and other observers in the southern hemisphere have enjoyed much success in photographic nova patrol work among the rich starfields in the direction of the Galaxy's centre. Paul Camilleri of Victoria, Australia has many discoveries to his credit. Patrols for novae in the Magellanic Clouds by Robert McNaught (an expatriate Scot) and Gordon Garradd from Australia have also been productive. Even photographs taken for purposes other than nova patrolling can be of value. Famously, the American amateur Ben Mayer unwittingly obtained pre-discovery images of Nova Cygni 1975 on the rise while attempting to capture meteors on film.

Table 10.1 lists novae discovered from 1975 to mid-1998. The annual discovery rate is quite variable, with none or only a single eruption in some years, up to the

Table 10.1. Novae since 1975

Year	Constellation	RA h	RA m	Dec	Peak Magnitude
1975	Sct	18	52.7	7° 47'	7.3
	Aql	19	15.5	0° 41.7'	11.5
	Sgr	17	55.3	–28° 22'	8.4
	Cyg	21	09.9	47° 57'	1.2
1976	Oph	18	00.9	11° 48'	8.8
	Vul	19	27.1	20° 22'	6.2
1977	Sag	19	37.0	18° 09'	7.0
	Sgr	18	35.2	–23° 23'	9.3
1978	Ser	17	49.0	–14° 43.1'	8.3
	Cyg	21	40.6	43° 48'	6.0
1980	Sgr	18	16.5	–24° 25'	9.0
1981	CrA	18	38.5	–37° 35'	7.0
1982	Aql	19	20.8	2° 23.6'	6.5
	Sgr	18	31.7	–26° 27'	8.5
1983	Sgr	18	04.7	–28° 49.9'	9.5
	Mus	11	49.6	–66° 57.7'	7.8
	Ser	17	53.0	–13° 58'	11.0
	Nor	16	09.9	–53° 11.5'	9.4
	Tri	02	42.2	32° 18.8'	15.0
1984	Vul	19	23.9	27° 16'	6.2
	Sgr	17	50.5	29° 01.5'	11.0
	Aql	19	14.1	3° 39'	10.0
	Vul	20	24.8	27° 40'	6.8
1985	Sco	17	53.3	–31° 49.2'	10.5
1986	Cen	12	17.7	–55° 34.9'	10.8
	Cyg	19	52.7	35° 33'	8.7
	Cen	14	32.2	–57° 24.5'	4.6
	And	23	09.5	47° 10'	6.3
1987	Sgr	17	56.5	–32° 16.2'	10.0
	Vul	19	02.1	21° 40'	7.0
1988	Oph	17	08.8	–29° 34.0'	8.5
1989	Sco	17	48.6	–32° 31'	9.4
	Sct	18	47.0	–6° 14.7'	8.5
1991	Her	18	44.3	12° 10.9'	5.4
	Oph	17	40.1	–20° 05.7'	9.3
	Cen	13	46.7	–62° 54.6'	8.7
	Oph	17	17.2	–26° 43.5'	10.5
	Sgr	18	10.9	–32° 14'	7.0
	Sct	18	44.4	–8° 24'	10.5
	Pup	09	09.7	–34° 58.5'	6.4
1992	Sgr	18	06.8	–25° 21'	7.0
	Cyg	20	30.5	52° 37.9'	5.0
	Sco	17	03.9	–43° 12'	8.2
	Sgr	18	20.3	–28° 23.7'	7.8
	Sgr	18	20.6	–23° 01.4'	9.0
1993	Oph	17	22.1	–23° 09'	9.5
	Aql	19	10.5	1° 28'	7.6

(continued overleaf)

Table 10.1. *Continued*

Year	Constellation	RA h	m	Dec	Peak Magnitude
	Sgr	18	09.7	–29° 30'	7.7
	Lup	14	28.4	–50° 7'	8.0
	Cas	23	39.4	57° 14.4'	5.1
1994	Oph	17	32.8	–19° 17.7'	7.8
1995	Cir	14	45.0	–6° 54.0'	7.4
	Aql	19	05.4	–1° 42.1'	7.5
	Cen	13	02.5	–60° 11.6'	7.6
	Cas	01	05	54° 02'	8.9
	Sgr	18	25.7	–18° 7.3'	10.0
1997	Sco	17	54.2	–30° 02'	8.5
1998	Sgr	18	21.6	27° 32'	7.8
1998	Oph	17	32.0	19° 14'	9.5

exceptional number of discoveries (seven) in 1991. Of the fifty-nine novae, only six reached mag. +6 or brighter, becoming theoretically visible to the naked eye in good conditions, and only one of these, Nova Cygni 1975, was really obvious. The denser starfields of the southern Milky Way are only slightly more productive of discoveries than the northern sky – it could be that some novae are missed in the richer background of the former.

It is impossible to predict when a nova to match the short-lived naked-eye spectacle of early autumn 1975 will next occur, but it is quite likely that the initial discovery will be made by an amateur observer carrying out a systematic visual or photographic patrol.

10.2 Supernovae

The most spectacular stellar cataclysms of all are supernovae, which mark the demise of particularly massive stars. Broadly, two types are recognised.

Type I supernovae are essentially conventional novae on a very much larger scale, consisting of binary systems in which a white dwarf accumulates matter from its red giant partner until its mass exceeds the Chandrasekhar limit of 1.4 times the Sun's mass. At this point there is massive nuclear runaway and the dwarf explodes, often resulting in the complete destruction of the system.

Type II supernovae are the end points in the lives of stars of more than 8 solar masses. These stars consume their nuclear fuel at a prodigious rate during their cosmologically short lifetimes, shining brilliantly for a few tens of millions of years before becoming bloated red supergiants. Ultimately, the nuclear burning of, in turn, ever-heavier elements in their interior leads to a near-instantaneous collapse of the outer layers, followed by a "rebound" which blows the star apart as a supernova. This fate awaits such beacons of our night sky as Rigel or Vega a few million years in the future. Orion's famous red supergiant Betelgeuse is further along in its life cycle; it too will meet its end in a Type II supernova explosion in a few million years' time.

Remnants of past supernovae can be observed as objects such as the Crab Nebula (M1) in Taurus, which marks the site of a stellar catastrophe witnessed by early astronomers in AD 1054, or the more extended Veil Nebula in Cygnus, the tattered remains and shock wave from a massive star which exploded probably 30 000 years ago.

In some supernova remnants, a neutron star or pulsar – the rapidly spinning core of the collapsed star – may be found; really massive stars may even undergo further collapse to form black holes. Exotic objects such as pulsars are of interest to professional astronomers, who are equipped to observe them. To many professionals, supernovae are the ultimate cosmological laboratories in which to test theories of nuclear physics. Some of the rarer elements can be produced only in the extreme conditions of a supernova explosion.

While we can identify those stars in our galactic neighbourhood which are likely candidates for supernovae in the distant future, such events are rare on human timescales. No supernova explosion has been recorded in our Galaxy since 1604, a few years before the telescope was first applied to astronomy. This almost 400-year dearth of nearby supernovae was a source of some consternation to astrophysicists until the welcome discovery of Supernova 1987A in the Large Magellanic Cloud on 1987 February 24. This Type II supernova has been the subject of close scrutiny ever since. Sooner or later it is to be hoped that a conventional nova patrol may catch the next Galactic supernova on the rise. Perhaps we may even enjoy a couple in the comparatively near future, if the same apparently perverse random statistics apply as for comets (Section 8.6).

The occurrence of supernovae in external galaxies, such as the nearby Large Magellanic Cloud, provides astronomers with a reasonable number to study from afar each year. The first recorded supernova beyond our Galaxy was the explosion, seen in 1885 August, of a star designated S Andromedae in the Andromeda Galaxy (M31); at its peak it reached mag. +6. Supernovae in more distant galaxies naturally appear a good deal fainter.

Supernovae in external galaxies are of importance to astronomers trying to refine the distance scale of the Universe. The absolute magnitudes of supernovae are fairly constant, so such objects can be used – like Cepheid variable stars in comparatively nearby galaxies – as "standard candles" whose apparent luminosity can be used to derive their distances.

In order to obtain a reasonable light curve and spectrum (from which the type can be determined), professional astronomers need early alerts to the appearance of supernovae. This is an area where, provided that care is taken to avoid issuing false alarms, the community of amateur astronomers can contribute – and *has* contributed – much of scientific value. Dedicated amateur observers have the time and ability to make systematic searches of suitable candidate galaxies to check for the occurrence of previously unreported supernova eruptions. While professional surveys, including some automated systems, catch a large proportion of new objects, the sheer number of galaxies to be patrolled means that some supernovae slip through the net, and can be found by suitably equipped amateurs.

10.2.1 Visual Searches

Before searching for supernovae in external galaxies, the galaxies themselves need to be found. Observers intent on a programme of visual supernova searching should therefore ideally spend some time learning the skills of deep-sky observing, and do so to a level beyond merely spotting the bright objects. Visual supernova searches entail the regular inspection of certain suitable candidate galaxies, checking for the appearance of new stars, often in their outer regions. Expertise in star-hopping the telescope (a minimum aperture of 250 mm is required) to the correct field is essential.

Good candidates for supernova searching are those spiral galaxies which are presented more or less face-on towards us. As anyone who has compared the face-on Triangulum Galaxy (M33) with its more edgewise neighbour M31 in Andromeda will know, such objects are often of low contrast with the sky background. This is not necessarily a problem: any supernova which erupts in one of the spiral arms, well out from the central nucleus of such a galaxy, should be obvious even if the background haze of unresolved stars in its neighbourhood cannot be seen.

More important is a knowledge of the starfield, made up of foreground stars in our own Galaxy, that surrounds a particular target. Once a list of suitable galaxies has been decided upon by the observer, reference sketches of their vicinities can be prepared from atlases such as *Uranometria 2000.0* or, better, from a combination of the atlas and the actual view through the telescope which is going to be used routinely for the patrol. Alternatively, the TA/BAA Patrol can supply ready-made charts for a reasonable fee.

Several sessions at the eyepiece will usually be required before the observer begins to become reasonably familiar with the field of each galaxy on the list, and it is best to start with a small programme and build up the numbers of objects patrolled as one becomes more experienced. Eventually, with practice, several galaxies can be checked in an hour's observing. Good candidates for supernova explosions are face-on galaxies with well-developed H II regions in their spiral arms: supernovae are commonest in such star-forming regions. Most photographic atlases of galaxies allow H II regions to be identified, and some time spent browsing through these may help in selecting suitable objects to patrol.

As in nova patrolling, there are of course pitfalls. Supernova suspects have to be checked for any motion over the course of an hour or so, just to ensure that they are not asteroids traversing the field. Variable stars in our Galaxy may also cause confusion, but should be readily identifiable from a good atlas.

The world's leading visual supernova searcher is the Australian observer the Reverend Robert Evans, who usually carries out his patrols with a 410-mm reflector. His intimate knowledge of hundreds of galaxy fields allows Evans to check about one field every minute on a good night, and at the time of writing he had 36 discoveries to his name.

10.2.2 Photographic and CCD Searches

When the suggestion was first made in the early 1980s that amateur astronomers could indeed usefully search for supernovae, the recording medium of choice was photographic emulsion. Patrol photography entailed long hours, carefully guiding exposures on fast film, each exposure having to be repeated as a precaution against flaws in the emulsion. The increased availability of personal computers and, in particular, sensitive CCD cameras has revolutionised many areas of amateur astronomy. Imaging faint objects has undoubtedly become easier (if not yet trivial), and at the leading edge of amateur CCD work there are many observers now taking images to rival those that were obtainable with the best professional equipment just two decades ago.

Taking the new technology beyond the acquisition of "pretty pictures", the supernova searchers have embraced it as a means of making patrol work faster, more efficient and more productive. Sensitive CCD cameras coupled to even quite modest telescopes can record faint stars in short integrated exposures, ending the need for special film treatments and long minutes of guiding at the finder telescope. Telescopes can even be driven, and programmed by suitable computer software to slew automatically to a series of objects, recording patrol images of each one. It is then a matter of checking on the computer screen for the appearance of any new stellar objects relative to the reference frame.

A collection of reference images should ideally be made using the patrol instrument. Increasingly, images of galaxies are becoming available on CD-ROM: the *Real Sky* CD-ROM set, for example, is a valuable resource (Section 11.4). Photographic atlases, such as the *Carnegie Atlas of Galaxies*, are also still useful reference sources.

CCD images for supernova patrolling do not have to be particularly deep, or show a lot of galactic detail. Indeed, it is preferable not to overexpose the galaxy, which would risk losing a supernova in 'burned out' brighter inner spiral arms. Many discovery images show little more than the brighter nuclear regions of the galaxy, and the supernova almost in isolation.

Using a CCD camera attached to a 200-mm off-the-shelf catadioptric reflector, an observer might hope to image and check 10–15 galaxies per hour, allowing for duplicate exposures. Each galaxy need be checked only once or twice a month, and in a good spell of clear weather up to 200 might be patrolled on moonless nights.

Since electronic imaging allows fainter objects to be recorded, much more caution is required in CCD supernova patrols. The average discovery magnitude for supernovae is around +14. In this region of the magnitude scale there are a great many asteroids which can be mistaken for supernovae, and rigorous checking for movement (over the course of an hour or so) and of the available asteroid databases (ease of access to which is another benefit of modern PC technology and the Internet) is essential. Only when the suspect is confirmed not to be an asteroid, or a known foreground variable star in our own Galaxy, should any attempt be made to report a discovery. The element of teamwork in observing networks such as the UK Nova/Supernova patrol is very useful for obtaining independent confirmation. In the light of recent embarrassing false alarms – which wasted valuable, limited professional telescope time – observers suspecting that they have made a supernova discovery are advised to have their findings checked via one of the intermediary amateur clearing houses, rather than reporting directly to the IAU Central Bureau.

Several observers around the world have now made supernova discoveries using CCD systems. Italian and Japanese observers have enjoyed a great deal of success. More recently, the apparent "jinx" on supernova work for observers in the UK was finally broken by the discovery of a supernova in the galaxy NGC 673 on the night of 1996 October 23 by Kent amateur Mark Armstrong. Further discoveries have followed, including another three for Armstrong (two of which are shown in Figures 10.2 and 10.3, *overleaf*), and one each for Stephen Laurie, Tom Boles and Ron Arbour. In the United States, Michael Schwartz of Oregon has made several discoveries.

Bright supernovae of recent years have included a number in some of the Messier Catalogue galaxies; for example one was found in M66, in Leo, in 1989. Supernova 1993J was discovered in M81 in Ursa Major by Francisco Garcia from Lugo, Spain, on 1993 March 28. This Type II object was the brightest northern-hemisphere supernova for over 35 years, reaching a peak of mag. +10.5, before fading to mag. +13 by June and mag. +15 by October.

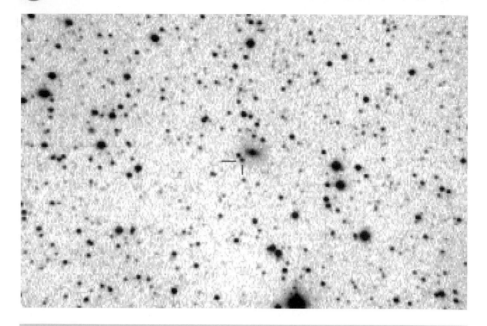

Figure 10.2.   Discovery image of the Type I Supernova 1998V, obtained by Mark Armstrong on 1998 March 9/10 using a CCD camera on his 200-mm reflector. At discovery, the supernova, in the galaxy NGC 6627, was at mag. +15.5.
Courtesy Mark Armstrong

Figure 10.3.   Discovery image of the supernova 1998aq in the galaxy NGC 3982 by Mark Armstrong. At discovery, the supernova was of mag. +14.9.
Courtesy Mark Armstrong.

Another visual discovery, Supernova 1994I erupted in the much observed Whirlpool Galaxy (M51). It was discovered by US amateurs Wayne Johnson and Douglas Millar on 1994 April 2 when at mag. +13.7. Supernova 1994I appeared quite close to the nucleus; it peaked at mag. +13.2 around a week after discovery, then faded to below mag. +15 before the month's end.

Both these objects were sufficiently bright for observers with reasonably large telescopes, using charts provided with the initial alert circulars by organisations such as *The Astronomer*, to obtain nightly magnitude estimates. These can be a useful supplement to professional photometric studies, and again emphasise the benefits of rapid communication.

References and Resources

Bryan, R and Thompson, L, *The Supernova Search Charts and Handbook*. Cambridge University Press (1989).

Evans, R, "Experiments in visual Supernova hunting with a large telescope" in *1998 Yearbook of Astronomy*, edited by P Moore. Macmillan (1998).

Murdin, P, *End in Fire*. Cambridge University Press (1990).

Ridpath, I, 'The man with the astronomical memory'. *New Scientist* 96 *1336* 719–21 (1982).

Sandage, A and Bedke, J, *The Carnegie Atlas of Galaxies*. Carnegie Institute (1994).

UK Nova/Supernova Patrol: Guy Hurst, 16 Westminster Close, Basingstoke, Hampshire, RG22 4PP, United Kingdom.

UK Nova/Supernova Patrol Web Page: http://www.demon.co.uk/astronomer/novae.html

Chapter 11

Information Sources

The successful observation of astronomical phenomena – transient or otherwise – usually depends on appropriate resources and preparation. No matter how large your telescope, or how high-tech the ancillary equipment, access to a suitable atlas is an essential. And while many transient phenomena are unpredictable, as the preceding chapters have shown, there are others such as eclipses, occultations or meteor showers for which at least approximate predictions can be given, allowing some degree of advance preparation.

Many of the most useful aids to successful observing can usually be obtained from the suppliers who sell equipment such as telescopes or binoculars. Most equipment retailers carry a stock of computer software and books. It should also be possible to order books through local bookstores, or through specialist booksellers who advertise in the popular magazines.

11.1 Atlases

The reference frame of right ascension and declination used by astronomers to define the positions of stars, planets and other objects is not fixed. As a result of the continuous precession of the Earth's axis of rotation, the first point of Aries, where the Sun moves from south to north of the celestial equator at the vernal equinox, marking the zero point of right ascension, creeps gradually westwards by just over 50″ each year. The long-term result is that whole the RA–dec. grid gradually shifts relative to the stars themselves. Rather than update the RA–dec. coordinate system annually, astronomers find it convenient to change at standard epochs every 50 years. So, for example, for epoch 1950.0 the position of Castor was given as RA 07^h 31.4^m, dec. +32° 00′, whereas its 2000.0 position is RA 07^h 34.6^m, dec. +31° 53′.

Until the late 1970s most of the widely used atlases were plotted for the epoch of 1950.0. More recently, epoch 2000.0 atlases have become very much the standard, and will remain so until, perhaps, the 2030s, when those based upon epoch 2050.0 will become more appropriate. Epoch 2000.0 positions have been the standard for Solar System objects – a decision of particular relevance for observers of asteroids and comets – since 1989.

A long-used basic standard for most amateur astronomers has been *Norton's Star Atlas and Reference Handbook*, published through nineteen editions up to 1998. The current version, with the RA–dec. grid adjusted to epoch 2000.0, is still very useful as a wide-field atlas. Its eight charts – plotted, as the name suggests, for epoch 2000.0 – show stars to the naked-eye limit of mag. +6.5. *Norton's* is a convenient general-purpose atlas whose charts are useful as a quick, approximate guide for locating objects, and the accompanying reference section provides useful information.

For more detailed work to a fainter magnitude limit, more specialised (and more expensive) atlases are required. Many observers swear by *Sky Atlas 2000.0*, which divides the sky into twenty-six large-scale colour-coded charts to a limiting magnitude of +8.0. Since its publication in 1989, *Uranometria 2000.0*, another large-scale atlas to a limiting magnitude of +9.0, has been in widespread use by comet and asteroid observers. *Uranometria 2000.0* comprises two volumes, covering northern and southern skies.

More expensive is the three-volume *Millennium Star Atlas*, based on the high-precision positional data obtained by the Hipparcos satellite. The charts in this atlas have a fainter limiting magnitude of +11.0, making them suitable for work on faint comets, asteroids and deep-sky objects.

11.2 Catalogues

Identification of the myriad stars and other objects shown in the deeper atlases is, obviously, important. For example, where a couple of "unknown" stars from the field have been used to obtain a magnitude estimate for a comet, their identification and actual magnitudes will need to be found in order to process the observational data. Volume 1 of *Sky Catalogue 2000.0*, a weighty tome published as a companion to *Sky Atlas 2000.0*, is a most useful resource in this respect, listing accurate positions, magnitudes and other details for over 50 000 stars to a magnitude limit of +8.0. Volume 2 of *Sky Catalogue 2000.0* gives data for deep-sky objects, as does *The Deep Sky Field Guide to Uranometria 2000.0*.

11.3 Ephemerides, Annual Handbooks and News Sources

Many transient phenomena, such as occultations and eclipses, are predictable, and ephemerides listing the expected time and geographical visibility are therefore available well in advance. Predictions can be found in regular annual publications. Lengthy tables of astronomical data are given in the professionals' standard source, *The Astronomical Almanac*, published jointly each year by the US Naval Observatory and the HM Nautical Almanac Office.

More specifically designed for the active amateur observer are the BAA *Handbook* and the RAS of Canada *Observer's Handbook*. Both these publications provide listings of positional and other information for the Sun, Moon and planets, as well as details of lunar occultations, meteor showers and other phenomena such as minima of the eclipsing binary Algol. These handbooks are invaluable to serious amateurs, and are issued as a membership benefit by the respective organisations; they can also be obtained separately from specialist astronomical book suppliers or equipment retailers. At the very least, the annual handbooks are helpful in planning observations so as to avoid moonlight, or in timing observing trips to dark locations to coincide with specific events.

Many more general guides appear each year, of which *The Times Night Sky* is one of the best known to the UK public. While fine as a rough guide for the casual skywatcher, none of these really competes with the RASC or BAA publications in providing the detailed information required by the more serious observer.

As we have seen, not all events are predictable. The arrival of a new bright comet, or the eruption of a nova, cannot be forecast or included in the annual publications, however good these may be. For knowledge of such events the observer must rely on information sources published on shorter timescales.

Astronomy is served by a number of monthly publications with a reasonably short lead-time before their appearance on the news-stands. The US magazines *Sky & Telescope* and *Astronomy* often carry reports of events which happened only a couple of months earlier, as does their UK counterpart *Astronomy Now*. Sometimes it is possible for these periodicals to give good advance notice of forthcoming events, as they did with the apparition of Comet Hale–Bopp in 1997. For events which move more rapidly – Comet Hyakutake in 1996, say, taking a few weeks from discovery to being visible at its best – even faster information sources are required.

Most of the major national amateur astronomical bodies issue news circulars carrying alerts of new discoveries of potential observational interest. The BAA *Circulars* (available as a supplementary subscription item) have long been a key source of such information for amateur observers in the UK. Alerts can be issued by such means within a week or so, sufficient for many phenomena. With the advent of e-mail, however, even this short timescale is now seen as tardy by observers who pursue phenomena such as supernovae. Electronic circulars issued by *The Astronomer* have become the standard for observers in a number of fields.

Enthusiasts in certain fields of work naturally keep in fairly close touch with one another by similar means. Possible supernova discoveries have been verified on same the night by e-mail contact between observers, for instance. Alerts to other transient astronomical phenomena such as auroral storms are often passed by telephone between members of local astronomy clubs. The means of rapid information relay available to the modern amateur astronomer are many and varied!

11.4 Software

An area which has really taken off in recent years has been the production of personal computer programs for astronomy. Some educational software certainly provides good value entertainment for cloudy nights, and there are a host of CD-ROMs with Hubble and Voyager images, and images from ground-based telescopes.

A very valuable resource for deep-sky observers and, particularly, supernova hunters, is the *Real Sky* set, available from the Astronomical Society of the Pacific. *Real Sky* is a digitised version, on CD-ROM (eight disks for the northern hemisphere, ten for the southern), of the National Geographic/Palomar Observatory Sky Survey and its southern hemisphere counterpart, the UK Schmidt Survey. The detailed, deep images are ideal for rapid checking of possible supernovae, asteroids or comets in the field.

Perhaps of most general use are good "desktop planetarium" programs, which can access digitised star catalogues (*Sky Catalogue 2000.0* is available in this form, for example, as is the *Hubble Guide Star Catalogue*) and show the appearance of the sky, or part of it, at times and locations that can be set by the user. It has to be said that some of the simpler programs are less useful (and a lot more expensive) than an old-fashioned planisphere. The advanced versions, however, are of great value to the serious observer. The days of exposing one's expensive, case-bound paper atlas to the hazards of the cold and damp night air are numbered. In future, observers are more likely to run off a print-out of the relevant field(s) for their target object(s), which can be scribbled upon, annotated, and discarded without major grief once it has served its purpose.

Among the useful CD-ROM desktop planetarium programs which include a great deal of additional material of educational interest is *Redshift* (now in version 3.0). This program can be used to generate star charts to a magnitude limit of +8.0, and to plot asteroid positions and run simulations of events such as eclipses. The Graystel *Star Atlas 2* also allows such simulations, and can be used to predict occultations, show planetary movements and plot star charts.

The large data-storage capacity of CD-ROMs makes them ideal media for computer-readable star catalogues, and many are exploited by programs such as *Guide 6.0* or *TheSky*, which can even be used to control suitably equipped and mounted telescopes. *TheSky* is available in a number of versions, to ever-deeper limiting magnitudes of +9.0 (259 000 stars), +11.0 (1.5 million stars) or +15.0 (19 million stars!).

The range and quality of the very best star-plotting software increases almost weekly, and the potential user is advised to read the literature carefully, and take stock of the reviews in the monthly astronomical magazines for the most up-to-date advice. The bottom line, as with a telescope or pair of binoculars, is that a good desktop planetarium is one that can actually be used reliably and conveniently to help in preparing, enjoying and getting the most out of your observations.

Most observers who take CCD images use standard software packages such as Adobe *Photoshop* or *Paintshop Pro* for further image processing (the 'digital darkroom' has arrived!). Until comparatively recently, obtaining positional measurements from photographic images was a laborious process, requiring the use of expensive, cumbersome plate-measuring machines which required a lot of room, and were rarely to be found outside professional observatories. The arrival of digital image processing has changed the situation very much for the better. CCD-generated pixel images can be imported into programs, such as *Astrometrica*, which use the digital nature of the raw data to produce high-accuracy coordinates in right ascension and declination for objects recorded, such as comets, asteroids and supernovae (Aguirre, 1997). By digitising photographs onto CD media (a service increasingly available from high-street photographic stores), meteor observers have also been able to speed up their data processing. Steve Evans of Towcester, Northamptonshire, has applied this technology very effectively to his studies of the Geminid meteor shower radiant and its movements.

Astrometrica works by simply aligning a cursor with, and clicking on, images on the screen. Digitised positions for known stars (from the computerised version of the *Hubble Guide Star Catalogue*) then allow the position of the "unknown" to be rapidly determined.

11.5 The Internet

Of all the aspects of computing to have revolutionised not just astronomy, but all fields of scientific endeavour, has been the arrival of the Internet, a global set of resources available for virtually instant inspection by anyone equipped with a personal computer linked to a modem. In common with software, the astronomical resources on the Internet vary greatly in quality, and ultimately it is up to the individual user to decide just which sites are worth regular inspection. By its very nature, the World Wide Web changes rapidly, and sites may appear and disappear over the longer term. Among the more reliable are those maintained by the major astronomy magazines, which often carry breaking news some weeks before it can appear in print. The active observer will find the site maintained by *The Astronomer* magazine a particularly valuable resource, with news of recent discoveries often posted within a matter of days. The BAA too maintains a good site, which can be used to plan occultation and other observations (Sections 5.2 and 7.1.1). Other appropriate sites are given at the end of the preceding chapters. Contact with local astronomical societies can often be made through the Internet; many proudly display recent observations or other achievements on their topical Web pages.

References and Resources

Aguirre, EL, 'Comets, asteroids, and *Astrometrica*'. *Sky & Telescope* 94 2 72–4 (1997).

Astrometrica: Herbert Raab, Schrammlstrasse 8, A-4050 Traun, Austria. Web Site: http://mars.planet.co.at/lag/astrometrica/astrometrica.html

Astronomy magazine, Kalmbach Publishing Co., 21027 Crossroads Circle, P.O. Box 1612, Waukesha, WI 53187, USA. Web site: http://kalmbach.com/astro/astronomy.html

Astronomy Now magazine, Pole Star Publications, P.O. Box 175, Tonbridge, Kent TN10 4ZY, UK. Web site:http://www.astronomynow.com

BAA Web Site: http://www.ast.cam.ac.uk/~baa

Cragin, M, Lucyk, J and Rappaport, B, *Field Guide to Uranometria 2000.0*. Willmann-Bell (1993).

Graystel Software Ltd, 175 Pershore Road, Evesham, Worcestershire WR11 0NB, UK.

Guide 6.0: Project Pluto, 168 Ridge Road, Bowdoinham, Maine 04008, USA.

Hirshfeld, A, Sinnott, RW and Ochsenbein, F, *Sky Catalogue 2000.0*, 2nd edition. Cambridge University Press (1991).

Real Sky CD: Astronomical Society of the Pacific, 390 Ashton Avenue, San Francisco, CA94112, USA.

Redshift 3.0: Piranha Interactive Publishing Inc., 1839 West Drake, Suite B, Tempe, AZ 85283, USA. Published in the

UK by DK Multimedia, 9 Henrietta Street, London WC2E 8PS.

Ridpath, I (editor), *Norton's Star Atlas and Reference Handbook*, 19th edition. Longman (1998).

Sinnott, RW and Perryman, MAC, *Millennium Star Atlas*. Sky Publishing/ESA (1998).

Sky & Telescope magazine, Sky Publishing Corporation, 49 Bay State Road, Cambridge, MA 02138, USA. Web site: http://www.skypub.com

SkyMap Pro: The Thompson Partnership, Lion Buildings, Market Place, Uttoxeter, Staffordshire ST14 8HZ, UK.

The Astronomer magazine: Peter Meadows, 6 Chelmerton Avenue, Great Baddow, Chelmsford, Essex, CM2 9RE. Web Site: http://www.demon.co.uk/astronomer

TheSky: Software Bisque, 912 Twelfth Street, Suite A, Golden, Colorado 80401, USA.

Tirion, W, *Sky Atlas 2000.0*, 2nd edition. Sky Publishing (1998).

Tirion, W, Rappaport, B and Lovi, G, *Uranometria 2000.0*. Willmann-Bell (1990).